D1228591

Limited by Design

Limited by Design

R&D Laboratories in the U.S. National Innovation System

Michael Crow and Barry Bozeman

COLUMBIA UNIVERSITY PRESS ◆ NEW YORK

Columbia University Press
Publishers Since 1893
New York   Chichester, West Sussex
Copyright © 1998 Columbia University Press
All rights reserved

Library of Congress Cataloging-in-Publication Data
Crow, Michael M.
    Limited by design : R&D laboratories in the U.S. national
innovation system / Michael Crow and Barry Bozeman.
        p.   cm.
    Includes bibliographical references and index.
    ISBN 0–231–10982–2 (alk. paper)
    1. Research, Industrial—United States—Laboratories.
    2. Technology and state—United States.   I. Bozeman, Barry.
    II. Title.
    T176.C76   1998
    338.97306—dc21   98–5204

⊚

Casebound editions of Columbia University Press books are printed on permanent and
durable acid-free paper.
Printed in the United States of America
c 10 9 8 7 6 5 4 3 2 1

*We dedicate this book to Stephen Loveless and Walter Meyer,
great friends and NCRDP colleagues who, all too quickly, went
on to better things.*

CONTENTS

TABLES

Research and development (R&D) laboratories are among the most complex, dynamic institutions yet invented. Public policies affecting R&D laboratories typically treat them as simple, homogeneous, and stable. In this book, we seek to identify some of the dimensions of U.S. R&D laboratories' complexity and to develop science and technology policy prescriptions attuned to the complexity of the national innovation system. Our book takes an "institutional design" perspective in analyzing R&D laboratories, a perspective grounded in systems-level organizational studies and characterized by a prescriptive orientation. While the book benefits from a great deal of data analysis, our approach is to step back from the data accumulated over many years and to speculate about implications for the design of the U.S. national innovation system.

Our ideas about R&D laboratories have been shaped by our experiences in the National Comparative Research and Development Project (NCRDP), one of the most comprehensive efforts to develop empirically based profiles of national innovation systems. Begun in 1984, the National Comparative Research and Development Project is an ongoing study of the technical enterprise in the United States and other industrial nations. More than thirty researchers from seven universities in four countries have been directly involved in the NCRDP, developing data, writing research reports, and helping to design new aspects of the project.

During the past fourteen years of working on the NCRDP, we have interviewed or received questionnaires from thousands of scientists, science administrators, and science policy makers, chiefly from the U.S. and Japan but also from Canada, Russia, Korea, Germany, Taiwan, and England. We have visited research and development (R&D) laboratories of every sector and stripe, industry, government, and university. Some of the more interesting labs we've visited were "hybrids" where public and private ownership and funding were so complicated as to defy classification by increasingly worn and threadbare categories. We have talked with researchers and administrators in many of the largest R&D laboratories in the world, such laboratories as Sandia National Laboratory in the U.S., Lucky Goldstar in Korea, the National Research Institute for Metals in Japan. But we've spent a good deal of time in the R&D hinterlands—the Fort Keough Livestock Research Center in Ft. Collins, Colorado, Chalk River Atomic Laboratory in rural Ontario, and the Brewing and Distilling Research Institute in Japan (which, incidentally, conducted our favorite laboratory tour).

The NCRDP has been sponsored by a diverse set of agencies including the National Science Foundation (the chief funding source), the Department of Commerce, the Department of Energy, the New York State Research and Development Authority, Syracuse University, Iowa State University, and— testimony to our earnestness—our own meager funds. At this point many have used NCRDP data, and we hope many more will (the data are in the public domain), but we have been there since the humble beginnings— Crow's self-funded dissertation (1985) on R&D labs.

In this book we reflect upon our experiences with the NCRDP, the extensive quantitative data gathered, but also the knowledge gained from lab visits and interviews, and suggest implications for science and technology policies and strategies. While we make extensive use of the data gathered from the NCRDP, this book is *not* a systematic synthesis of research findings. Instead of simply reporting the results of NCRDP studies, we decided to write a book that would use our data and experiences but would be more focused on the prescriptive implications flowing from the work. That is, now that we have spent a good amount of time learning about the operation, structure, and policies of R&D laboratories, what are the implications for science and technology policy? This focus fits nicely with one of our core biases—that public policies affecting R&D laboratories seem to pay little heed to the laboratories themselves. R&D laboratories are so often the "means" by which policy makers seek to accomplish so many important and diverse policy "ends" that we are continually amazed that policies pay so little heed to the character and quality of their instrument. Most policies either make no distinctions whatsoever—a laboratory is a laboratory— or they proceed on only the grossest of categories such as government v. industry, defense v. civilian, large v. small. The notion that some labs might do one thing well and others might do something else well—which seems obvious to us and to many of our colleagues—is almost never reflected in public policy.

After we had produced our standard academic goods (see appendices), we felt that we had learned things not easily communicated in the conventional, conservative, academic media. In our fifty or so presentations from the

NCRDP, our focus was usually on the data and explaining R&D laboratories' behaviors and structures. But, invariably, when the post-presentation discussions turned to policy controversies, we found not only that we had something to say but that the knowledge we had developed from studying R&D labs had strongly colored our policy ideas and prescriptions. Which is to say, we have opinions. This book airs them out.

We have studied R&D laboratories as a "system." That is, we have not only examined particular labs with their unique strengths and weaknesses, we have from the beginning taken as our fundamental premise the need to study the "system," the need to understand the significance of the whole to the part and of the part to the whole. This system perspective may be the most significant contribution of our research. This may be surprising because one would expect that the elements of the U.S. R&D laboratory system would be well known. Such is not the case.

In examining the R&D system, rather than individual laboratories, sectors, or industries, national strategic questions quickly suggest themselves. How do the activities of labs fit together? Are some types of labs particularly suited for certain activities? Which labs are the strongest contributors to national innovation and productivity goals? How has the system been evolving? What impact does government policy have on different parts of the system? This last question is particularly compelling because it has become clear to us that most government policies are blunt instruments and few seem to have any idea of the depths, much less the direction, of their effects throughout the system.

Objectives

Our overall goal is to draw from the past fourteen years' experience profiling the U.S. (and other nations') R&D laboratories and to shed some light on pressing science and technology policy issues. The objectives of this book include:

1. to provide information about the role of the federal laboratories within the U.S. national innovation system;
2. to present ways of thinking about R&D laboratories that complement the simple (but still useful) categories such as government-industry-university;
3. to provide an overview of the commercial activities of federal laboratories and factors related to the effectiveness of those activities;
4. to provide data-informed recommendations for a national science and technology policy pertaining to the deployment of federal laboratories.

In pursuit of these objectives, we employ the results of the NCRDP, from the very first research results (Crow, 1985) to the most recent findings (Bozeman and Rogers, in press). We develop a number of themes, but the most prominent theme is that *"one size fits all" R&D policy, prominent for years in virtually every aspect of U.S. R&D policy, makes no sense.* The R&D laboratories of the U.S. and their foreign competitors are extraordinarily diverse. Treating laboratories as undifferentiated producers of undifferentiated technical goods undermines productivity and gives rise to inefficiency and waste. Understanding the diversity of R&D laboratories is not an easy task, but it may be prerequisite for successful R&D policies.

Since federal R&D policy can have the most direct and immediate impact on federal laboratories, and since federal laboratories have now been required by policy makers to play a growing role in U.S. economic "competitiveness," we give particular attention to the federal components of the U.S. R&D policy system. However, the role of federal laboratories cannot be considered in a vacuum. Their successful functioning depends, in large measure, upon complementing activities in the industrial, university, and nonprofit sectors. Thus, our focus is on the system, and the federal laboratories' roles within the system of R&D providers.

Acknowledgments

Since this book draws from a project vast in scope, the number of contributors meriting acknowledgment is well beyond the usual (if there is a "usual") number. Let us begin by saying that special thanks are owed to the many researchers who have taken part in the NCRDP, almost all of whom are listed as authors in the appendices and the bibliography. A second large group of contributors includes the thousands of busy laboratory directors and others who filled out questionnaires for us or granted us face-to-face or telephone interviews. We are particularly indebted to the hundreds of persons working in R&D labs who have hosted our site visits, in every case with patience and goodwill.

Studying labs is not cheap. We are grateful to the many funding agencies who agreed with us about the need for more empirical data on R&D labs. We are grateful for the support of the National Science Foundation (including grants from the International Division, the Research on Research program, and the Science Indicators Unit), the Department of Energy (especially the Basic Energy Sciences Division, which supports the current phase of the NCRDP), the Department of Commerce, Sandia National Laboratory, the New York State Energy Research and Development Authority, and Japan's National Institute of Science and Technology Policy. While not directly funding the NCRDP, the U.S. General Accounting Office, particularly Gerald Dillingham, has helped enormously by sharing data.

Several people who have contributed to the design and intellectual evolution of the NCRDP bear particular mention. Those who shaped the NCRDP in fundamental ways at its very beginnings include Albert Link, Stuart Bretschneider, and the late Stephen Loveless. Maria Papadakis is chief among the intellectual contributors to the later stages of the NCRDP. She played a major role in framing, executing, and analyzing NCRDP research on technology commercialization at U.S. federal labs, and she was a coprincipal investigator in NCRDP projects performed in Japan.

The magnitude of the NCRDP would not have been possible without the assistance of a great many highly capable alumni of Syracuse University's Center for Technology and Information Policy, some of whom wrote dissertations from NCRDP data, some of whom conducted interviews, some of whom worked tirelessly on establishing and maintaining quality data, all of whom have gone on elsewhere to enjoy distinguished careers. Thus, we are grateful to Karen Coker, Carol Cimitile, Gordon Kingsley, Dennis Wittmer, Maureen O'Neill Fellows, Mary Tyskiewickz, Sanjay Pandey, Lan Zhidong, Man Cho, John Mortensen, Sander Glick, David Coursey, Dianne Rahm, Phil Nowicki, Younghoon Choi, Heungsuk Choi, Julia Melkers, Rick Shangraw, and, particularly, the nonpareil Monica Gaughan.

At Georgia Tech, Bozeman's colleagues in the Technology Policy and Assessment Center and the School of Public Policy provided an excellent sounding board for research. This outstanding group of science and technology policy scholars includes David Roessner, Hans Klein, Juan Rodgers, Jan Youtie, Gordon Kingsley, Phil Shapira, Richard Barke, and Alan Porter.

At Columbia University, Chris Tucker played a vital role in the project and, indeed, is a coauthor of chapter 2 with Bhaven Sampat, a graduate student in economics at Columbia. Yevgenia Lyublinskaya provided her keen eye to the editing of our charts and figures, with a smile.

This book, and the NCRDP generally, required much staff support. We are grateful to Pat Simone, Jo Sheridan, Pat Knight-Williams, Chris Everson, and Rita Davis.

Naturally, the agencies funding our studies do not necessarily agree with our interpretations and hold no blame for any flaws or limitations of our data. The opinions expressed in our book belong to us alone, and indeed, we know for sure that they do not reflect the views of some of the people we acknowledge above. In fourteen years' collaboration many have made their disagreements abundantly clear.

One of the major points of this study is that our knowledge of R&D performing organizations is scant. Both a reason for and a result of this is that we do not have an intuitive sense of the range and diversity of laboratories in the American national innovation system. Although we present our data in order to make this point abundantly clear, we thought that it would be suggestive to provide some additional "content" with the purpose of broadening the popular understanding of R&D laboratories. Only minor editorial changes have been made to depersonalize this additional content.

Our solution was what we have come to call "lab windows." Included in the text are 50 laboratory windows from over 45 labs. These labs are vastly different in type, ranging from the Citrus Research and Education Center of the University of Florida's Institute of Food and Agricultural Sciences to the Los Alamos National Laboratory with 6,800 Univesity of California employees plus approximately 2,800 contractor personnel and annual budget of approximately $1.2 billion. We have attempted to provide a healthy cross-section of university, industry, and government labs, demonstrating that even the most common labs tend to defy this common sectoral typology. The content has been pulled from worldwide web pages developed by these laboratories and their parent organizations. We thought this information would allow these organization to demonstrate how they present themselves to their constituencies, stakeholders, and clients while also giving readers some beachheads for their own explorations into the vast population of organizations performing R&D. This information was collected from these organization's websites between March and June of 1997. As internet time is at least twice as fast as real time, this information will no doubt change rapidly. Yet, the information and the related web addresses will provide the reader with fruitful starting points.

Permissions were sought from the following labs:

Lawrence Livermore
Information Technology Lab
NIST
Applied Physics Lab, Johns Hopkins
Brookhaven National Laboratory
IBM Watson Research Center
Iowa Agriculture and Home Economics Experiment Station
UF/IFAS Citrus Research and Education Center
Design Engineering Research Center
AT&T Bell Labs
National Seed Storage Lab, USDA
Ames Laboratory, DOE, IA
Lucent Technologies
USGS
NIH

NACA
Lincoln Labs, MIT
David Sarnoff Laboratory
Los Alamos National Laboratory
Notre Dame Radiation Laboratory
Oak Ridge National Laboratory
SEMATECH
Carnegie Institution of Washington
MCNC
Minnesota Agricultural Experiment Station
Goddard Space Flight Center, NASA
Lawrence Berkeley National Laboratory
Stanford Linear Accelerator
Phillips Petroleum Laboratory
Battelle Memorial Research Institute
GTE Laboratory
Naval Research Lab
Texaco
Pfizer Pharmaceuticals
Eastman Chemical Company
Magnavox Advanced Product Lab
Allison Turbine Division Labs
Motorola Space and Systems Technology Group
Center for Applied Energy Research, University of Kentucky
IBM
Kitt Peak Observatory, NSF
PPPL
Argonne National Laboratory
Human Exposure Research Division, EPA
Army Cold Regions Research and Engineering Lab
Pittsburgh Energy Tech Center
The Federal Laboratory Consortium for Technology Transfer
Rome Laboratory, Air Force

Our study demonstrates that R&D organizations operate in complex environments where their sector of ownership may not closely correspond to the influences exerted upon them. We ask readers to take this into consideration when viewing the content in these lab windows. One example of the complexity was provided by representatives of the Stanford Linear Accelerator when they asked for the following statement to follow their web content: "Copyright for SLAC web pages belongs to Stanford University, which operates the Stanford Linear Accelerator for the United States Department of Energy." We hope readers will respect the copyrights of these organizations, as the labs seek to provide public access while maintaining the restraints of their own particular legal statuses.

Limited by Design

CHAPTER ONE

The Sixteen Thousand

Policy Analysis, R&D Laboratories, and the National Innovation System

Introductory Vignettes

June 1991, Los Alamos, New Mexico. Resting on 43 square miles of mesas and canyons in northern New Mexico is a vast set of concrete and tin structures, not a mammoth ski resort, but a research facility unique in the contemporary world. Los Alamos National Laboratory, the nest that in the 1940s incubated the world's most terrible weapon, is a sort of bizarre national monument to the power of organized science. Los Alamos has captured the public imagination. What better evidence than its having served as a set for a movie about the Manhattan Project (*Fat Man and Little Boy*, starring Paul Newman, no less, as General Leslie Groves, military director of the project).

We have a fleeting feeling that we are on a Hollywood set and expect to see planks propping up the facades. But this is the real thing, the workplace for more than 7,000 University of California employees and 3,500 contractor personnel, largely scientists and technicians. The contemporary Los Alamos remains a leader in the science and technology of weapons production, but it is also a contributor to advances in fundamental science, especially materials science and physics. This enclave is now also playing an important role in the region's attempt to develop and exploit *commercial* technology. A technology transfer agent in the Office of Research and Technical Applications explains to us that Los Alamos is committed to a lively program of technology transfer and support of industry, as evidenced by the more than a hundred licenses granted last year, by its more than 20 patents, and by the 34 cooperative research agreements. Its activities have ranged from helping to start new businesses to researching solutions to the nation's nuclear waste problem (at the same time Los Alamos also contributes to that very problem).

Los Alamos National Laboratory, a behemoth employing more doctoral-level physicists than can be found in all but a few nations, is many things.

Its unique scientific facilities and human capital permit it to make contributions to the world's stock of public domain scientific knowledge. Notwithstanding the extensive fundamental scientific research at Los Alamos, despite its growing commercial focus, it is still a weapons lab. Most of its budget goes to its nuclear programs. Today's Los Alamos would barely be recognizable to Oppie or to Fat Boy. Today's Los Alamos is a mix of diverse enterprises. Using its vast resources, it is going in many different directions at once. Perhaps too many directions. What is a government research and development (R&D) lab? Surely Los Alamos is one. But what is Los Alamos? It is an institution undergoing rapid change, and whatever it is, it's much different from what it was even only five years ago.

May 1987, Vermillion, South Dakota. The population of Vermillion, South Dakota, is about the same as that of Los Alamos, New Mexico. But the percentage of scientists and engineers in Vermillion is quite a bit lower, the percentage of farmers quite a bit higher. Vermillion is the home of the South Dakota State Geological Survey, a government laboratory of the state of South Dakota. A faint scent of what we think is alfalfa wafts in the morning breeze.

Unlike Los Alamos, no one has made a movie about the Vermillion lab, not even a single Public Broadcasting special. No visits by method actors trying to immerse themselves in the ambiance, treading the same hills as Oppenheimer. No Paul Newman. The Geological Survey has not, so to speak, exploded upon the world of science. Not that the scientists here are without ambition. Rather, they are without funds and personnel. Employing only 12 scientists, the annual budget at Vermillion is less than what is spent at Los Alamos in three hours. Worse, the lab director explains to us that he has received no new funding in *15 years*. He described in detail how he and his staff had built from scratch a handmade spectrometer—the alternative was to do without one. The South Dakota State Geological Survey, despite its obvious limitations, is an R&D lab: the scientists and engineers employed in this unit are actively engaged in research, expanding the boundaries of knowledge to the extent that their minuscule resources will permit them.

June 1992, Tokyo, Japan. Halfway around the world on a side street in suburban Tokyo sits the central headquarters of the National Research Institute for Metals. Unimposing, NRIM is housed in a set of dingy gray, pre–World War II era buildings (newer facilities in Tsukuba City were completed in 1995). Only the guard house and fence mark it as different from other neighborhood buildings. We remind ourselves that many of the scientific and technological advances that propelled the Japanese steel industry to world prominence in the 1950s and 1960s came from this modest physical plant.

In a government-issue conference room looking out on a beautiful garden, the director general of the laboratory explains to us that the role of NRIM has evolved considerably as the R&D capabilities of the Japanese steel and metals industries have developed world-class research facilities of their own. NRIM, which cut its teeth on applied research and technical assistance to industry, is refocusing its mission to fundamental research. Later, the leader of a major research division joins us, a self-confident and contentious scientist eager to get back to "more pressing business." He is clearly a cookie-cutter scientist, interested in publications and scientific prestige. We ask him why he gets involved in cooperative research with industry and he responds, "It's my duty, I'm paid by tax dollars." He hastens to add, "but I don't work with them if I don't like their personality."

March 1990, Hoboken, New Jersey. Stevens Institute of Technology is in Hoboken, and our appointment with the director of its polymer processing institute surprises us from the beginning. The title "director" is nearly universal in the U.S., just as "director general" is the main official title in Japan. But at Stevens the lab director is "president." This does not seem a bad name. We ask the president of the lab how much the lab receives in university funds. "None." How much in government funds? "None." Every cent of the lab's research budget comes from regional companies for whom it provides various software solutions and technical assistance.

After the interview we go across the street to McDonald's for a Big Mac. The manager of the restaurant is packing a gun. Obviously, the people working in this lab are in a tough environment—in more ways than one. This is an R&D laboratory but not most people's stereotype of a university lab. No ivy-covered buildings. Unsafe funding. Unsafe, period.

June 1992, Port Hueneme, California. "Fighting Seabees" reads the sign outside the laboratory. Inside the lab civilian engineers and scientists are working to develop new construction materials for rapid airfield deployment or naval pier construction, while at the same time developing enhancements for the construction of Navy housing. An internal Navy shop, this lab for the Navy's Construction Battalion has top-ranked engineers focusing on problems unique to the Seabees' "can do" motto, problems such as building concrete-based underwater structures or developing materials to reduce ocean-borne corrosion. This work goes on in the midst of Seabee headquarters. The researchers in this lab support the Seabees in the bulldozer and the cement truck and seem to have the same "can do" attitude that has long characterized the Seabees. They are problem-solvers first, scientists and technicians second.

June 1991, Ames, Iowa. The smell of cow manure isn't what one expects driving up to a major research facility but this is what confronts us at the National Animal Disease Center, a 500-person research institute. This 1950s vintage research lab in the cornfields is surrounded by scores of "outbuildings" where livestock of every variety are isolated for biomedical research. In 1991 the big issues seem to be how to take advantage of breakthroughs in retrovirus research in animals to inform AIDS research and how to best dispose of the dead animals resulting from the research. We note in passing that the farmers surrounding this lab probably make hospitable neighbors, not an animal rights advocate to be found for many square miles.

Historically, the lab has found cures for livestock cancers and played a major role in developing new veterinary techniques. The movement to linkages to human health is more recent. But the new research thrusts coexist with the old ones—seeking an AIDS vaccine alongside working on medicines to alleviate pig diseases. Speaking with the researchers, we find no hint that problems of porcine digestion are viewed as any less exciting than frontier research on human pharmacology.

April 1993, Rome, New York. Defense conversion in action. We are being treated to a presentation of the commercial activities of one of the more nontraditional Department of Defense laboratories, the Air Force's Rome Laboratory. It begins with the colonel's recently stated dictum, displayed prominently on an overhead transparency, "I had rather take the reins of a wild horse than beat a dead one." The wild horse is technology transfer and dual use technology, the dead one is, we suppose, left to the imagination.

Until very recently Rome Laboratory's external relations focused on "technology transition"(in Air Force parlance), transfer of products and processes into Air Force use. But the DOD world has changed. During the last month, Rome Laboratory has gone from one person devoted part-time to technology transfer to a full-scale Office of Research and Technology Applications (ORTA), employing three persons full-time as well as four "sorta-ORTAs," persons from research divisions working part-time on technology transfer. Perhaps coincidentally, Griffiss Airbase, the home of Rome Laboratory, has been targeted by the base-closing commission for significant downsizing.

Postscript, 1995: Griffiss Airbase is no more and Rome Laboratory, which survived three years after Griffiss, remains active in 1997 after significant political maneuvering by the New York congressional delegation.

April 1996, La Jolla, California. From the director's office the dominant sound is the ocean. From the sliding glass door opening onto the patio, the ocean is alive with surfers and the occasional dolphin. With a large seaside campus in trendy La Jolla, California, the Scripps Institution of Oceanography has grown into the largest center in the world for the basic science of the earth and how it works. With oceangoing research vessels, research divisions in climate, oceanography, geosciences, and marine biology, institutes in arctic and antarctic science, space, planetary physics and earth observations, Scripps has evolved into a whole earth or earth systems research institution that covers it all.

. The evolution of Scripps from a single stand-alone oceanographic research group in 1903 to an institution of 300 groups with 1,200 employees that also served as the seed for an entire university (University of California at San Diego) is a story about the centrality of scientific excellence at Scripps and field growth in the earth sciences.

With strong private interest and funding at its founding and with the independent spirit nurtured in the independence of a stand-alone campus and from a culture built around the sea, Scripps has evolved into an R&D laboratory that sets the course for research about the earth and has defined the development of the notion that humans are responsible for the management of the planet.

That belief and the science that supports it is in the air as one walks the campus, looks to the sea, and thinks about the future.

June 1997, Oracle, Arizona. From 1991–1995 Biosphere 2, this large "laboratory" of the earth in the desert, was ridiculed by the science community as bad science. Bad science because it literally came out of nowhere. No peer review of the concept, no peer review of the participants, and no oversight of the "science" being conducted. The result was rebuke and disassociation.

By June of 1997 the laboratory, now under the leadership of a major research university and with a manager straight out of the National Science Foundation was on the slow road to redemption. With a unique facility and a dream for earth systems science, the Biosphere 2 laboratory is undergoing an intellectual reconstruction. The director, with private and university resources, is building his own peer review mechanisms, building a community of scientists for the laboratory, and planting the seed for the culture of American science.

Even with unique research facilities, complex research questions, good scientists, and start-up resources, the director still struggles. He needs acceptance and understanding from the science community as much as anything else.

The Mysterious 16,000

During several years studying R&D laboratories, we have mused about why R&D policy makers pay so little attention to the institutions charged with carrying out their policy directives. Our experience tells us that R&D laboratories are among the most diverse sets of organizations one will ever find. Labs have unique and deeply entrenched cultures. Labs differ in technical capabilities, organizational structure, resources, market interaction, political and bureaucratic environment, and adaptability. We have not stumbled onto a reality unknown to others. Most observers recognize that labs have singular strengths and comparative advantages that need to be understood and exploited. Research and development (R&D) laboratories are so often the means by which policy makers seek to accomplish diverse policy ends, one might expect them to pay greater heed to the character and quality of the instrument. Yet, despite the widespread realization that R&D labs are quite complex, gross and undifferentiated policies remain the norm. Policies view labs as government or industrial, defense or civilian, large or small, but almost never in all their richness and diversity. Why? Science and technology policy emerges from a shroud of empirical darkness.

Several forces contribute to the rough-hewn quality of public policy for R&D labs, including political compromise, obsession with the budgetary aspects of policy, and the sustaining power of myths (e.g., the World War II legacy). But one reason for the blunt policy instruments is fundamental: the complexity of U.S. R&D laboratories is so daunting as to discourage attention to detail. There are approximately 700 federal labs directly funded by the U.S. government, at least 100 of which are of sufficient size and scope to be considered significant contributors in themselves to the national innovation system (NIS). Hundreds of university labs produce everything from new Ph.D.'s in chemistry to new synthetic proteins to new rat poisons. To this mix one must add nearly 14,000 industrial laboratories, about 1,000 of which are consequential to the national science and technology effort, including several leading-edge innovators. The sheer numbers defy an "up close and personal" perspective.

The avoidance of complexity and the unfamiliarity with the "systems" characteristics of U.S. R&D laboratories are certainly understandable. Generally, policy makers seek simplicity, attempting to develop rules of thumb and decision heuristics (Braybrooke and Lindblom, 1963). But in the case of R&D policy the results of oversimplification include ill-formed, poorly rationalized science and technology policies. These simplifications often have dire effects on the 700 federal labs, deployable assets that consumed almost 35% of the total federal R&D investment (of more than $70 billion) in 1996.

Problems in the analysis of R&D laboratories are easily understood, but not so easily remedied. The problems:

- analysis is based on limited information and, often, the wrong sort of information;
- analysis is based on just a few simplistic concepts and categories;
- analysis is based on the belief that ownership (government, industry, university) determines behavior.

The remedies:

- better information;
- more powerful and useful concepts and classification tools;
- a willingness to use information in analysis and evaluation of U.S. R&D laboratories.

The remedies are more difficult than they seem. The U.S. R&D laboratory system is remarkably complex, with more than 16,000 extremely diverse knowledge-producing institutions influenced by a congeries of rapidly changing environmental forces, many poorly understood. Gathering useful information is a costly and otherwise formidable task. Likewise, developing richer concepts and categories for laboratories is challenging. Where does one begin? Labs vary in so many significant dimensions—technology, creativity, organization, resources, sector, culture, market orientation—there is no obvious point of analytical departure.

The complexity in institutional setting, environments, and organizational characteristics of knowledge-producing organizations should be given greater emphasis in R&D policy. That is why we have written this book—to provide additional knowledge about the U.S. R&D laboratory *system*, the principal knowledge-producing units, the "factories" of the larger and more complex U.S. national innovation system. Our purpose is to suggest why it is important to think in terms of the system, to indicate ways to think about the system, and to show the relationship of system knowledge to R&D policy controversies. We begin by delving more deeply into the issue of why we continue to know so little about U.S. R&D laboratories while they are universally recognized as vital to the nation's capacity to generate science, technology, and economic and social progress.

WHY THE MYSTERY?

Ignorance of the U.S. system of R&D laboratories is certainly easy enough to document. Most people, even well-informed specialists, have no idea even about the number of U.S. R&D laboratories. On several occasions we have quizzed such people. Speaking with lab directors, scientists, congressional staff, employees of agencies with science and technology oversight functions (e.g., GAO, OMB), and policy analysts, we have simply asked: "How many R&D labs are there in the U.S., if we define a lab as focusing on science or engineering and having at least 25 full-time employees?" The answers range from a low of 500 to a high of about 20,000; the mean response seems to be about 3,000. (The best evidence we have developed shows about 16,000 to 17,000, defining an R&D laboratory as focusing on science or engineering and employing at least 25 personnel.)

When it comes to more specific questions about the U.S. R&D laboratory system, almost everyone is lost. Let's consider a few examples. Policy makers who feel that federal laboratories are fountains of technology that, after being unclogged by new policies, can spew forth their bounty to eager private sector firms, may wish to consider the present rate of technology licensing. A detailed review of the technology transfer outputs of federal labs in the mid-1990s indicated that while funding for these organizations was about $25 billion annually, the total number of new licensed technologies was only about 300 per year with very little economic return (less than $10 million annual) to the labs themselves (U.S. Department of Energy 1994b, 1996). Nor is there any systematic knowledge as to which federal laboratories are better at commercializing technology. Where lies the evidence that private sector firms are queued up to obtain federal laboratory technology?

There has been much confusion as to the very purpose of many of these federal laboratories. Consider the dilemma of Dr. Bruce Tarter, director of the Lawrence Livermore National Laboratory (LLNL), who in congressional testimony in March 1996 indicated that his efforts to transform the lab into an economic development force had to be blended with the lab's responsibilities for maintaining a safe nuclear weapons stockpile.

Lab Window 1.1: "LLNL: Conflicting Missions"

Lawrence Livermore National Laboratory
http://www.llnl.gov

Livermore is changing to meet today's challenges. By matching areas of expertise to pressing national and global challenges, Livermore is focusing on three areas of long-term importance where its contributions are unique and valuable:

I. Global Security: Reducing the Nuclear Danger

II. Global Ecology: Harmonizing the Economy with the Environment

III. Bioscience: The New Frontier

Beyond these primary foci, Livermore will continue to support other innovative science and technology initiatives that have the potential for high impact in their field and that reinforce the lab's scientific and technological strengths. In addition, Livermore remains committed to fostering science and math education to help ensure the scientific literacy of the general population and to inspire future generations of scientists and engineers.

Livermore's vision for the future aligns with the business areas identified in the Department of Energy's new strategic plan: national security, energy resources, environmental quality, and industrial competitiveness—all addressed through science and technology. The lab will build on and enhance partnerships with departmental staff to ensure excellence in the achievement of common goals.

Attaining these goals will also require new forms of cooperation among the national laboratories, universities, and industry. Key national facilities will be built and used by multi-institutional teams of researchers. Laboratory sites will be readily accessible to outside partners. The commercialization of new technologies will be the planned end product of these collaborative projects.

This vision for the future of the Livermore Laboratory is one of sustained, results-oriented excellence. Livermore is committed to serving the country as a national resource of scientific and technological expertise, dedicated to global security, the environment, and the future scientific needs of the nation.

- Integrating Multiple Disciplines to Solve Complex Problems
- Partnering as a Way of Doing Business
- Managing Effectively in a Changing Business Environment
- Focus and Change

How exactly should LLNL work to transfer new technology to start-up companies in Northern California and be a center for new technology-based business while it is trying to manage a $400 million per year effort that includes enhanced nuclear stockpile surveillance, revalidation process development through advanced computing, and agile manufacturing for robotic management of aging nuclear weapons? Doable, perhaps, but not easily.

Knowledge of the U.S. R&D laboratory system is important not only for "high policy" but also for laboratory-level policy. For example, most laboratory directors probably feel that it is a good idea to keep a lean administrative staff so as to devote as many resources as possible to scientific and technical personnel and their work. But what is a *good* administrative-technical staff ratio? What is an *average* administrative-technical staff ratio? No one knows.

In our view, there are two chief reasons for limited understanding of the R&D laboratory system. First, the system is so large and complex that it is extremely difficult to develop more than a superficial knowledge of the system or large components of the system. Second, and less obvious, the policy frameworks that dominate R&D policy work against system-level knowledge.

Nobody Fully Understands More Than a Few Labs

Knowledge of R&D labs comes in clusters and clumps or, more often, in singletons. Some people who work in R&D labs know a great deal about the lab in which they work. Our experience, however, indicates that most people who work in labs actually know very little even about their own lab. Aside from the director, a few key science administrators, and the occasional cosmopolitan scientist, most scientists know a great deal about their own research and very little about the lab, its missions, its connections to the outside world, its funding, budgeting and planning processes. Researchers know much more about other researchers who work on similar problems but on the other side of the world than they know about research in unrelated fields in the next building. Most scientists, except for their involvement in writing grant proposals, seem to assume mystical research funding processes. Money keeps coming, maybe not at the rate they wish, maybe with some seasonal variation, but it keeps coming. Most lab scientists give no more thought to rationalizing continued funding in terms of return-on-investment than a concert pianist gives to the cost-benefit analysis of piano sonatas.

Still, some people do have a thorough understanding of their own labs. They know what makes them tick, what the labs do well and not so well, who are the productive scientists, and who are the dogs who lap up resources and produce little. Some not only know their laboratories but have a fair knowledge of the environment in which their laboratory exists. But few understand the laboratories, the laboratories' environments and linkages to other laboratories. These linkages among networks of laboratories become more important each year. For example, the network built around the Information Technology Laboratory of the National Institutes for Standards and Technology is a perfect example. Here a federal lab plays an essential role, as defined by industry, as a network-centered "competent neutral asset" (Brownstein, 1996). In this case, industry wants standards, testing, neutral research, and evaluation of academic advances. The NIST lab provides this and centers the network.

Lab Window 1.2: "R&D Networks"

Information Technology Lab at NIST
http://www.nist.gov/itldiv.htm

The Information Technology Laboratory (ITL) responds to industry and user needs for objective, neutral tests for information technology. These are the enabling tools that help companies produce the next generation of products and services and that help industries and individuals use these complex products and services. ITL works with industry, research, and government organizations to develop and demonstrate tests, test methods, reference data, proof of concept implementations, and other infrastructure technologies. Because measurements create a common language for technology advancement, tools developed by ITL provide impartial ways of measuring information technology products so that developers and users can evaluate how products perform and assess their quality based on objective criteria. Tests also supply neutral ways to demonstrate that products satisfy required functions and features, work with other products, and conform to a specification or standard. ITL's activities support the development and use of information technology systems that are usable, scalable, interoperable, and secure. Program activities include: high performance computing and communications systems; emerging network technologies; access to, exchange, and retrieval of complex information; computational and statistical methods; information security; and testing tools and methods to improve the quality of software. We invite you to explore the ITL web site for in-depth information about these activities.

Directing a major laboratory requires a good understanding of environmental forces. A much smaller set of people seems to have considerable knowledge of constellations of laboratories. Some can reveal a great deal about laboratories working on pharmaceuticals, some know much about aerospace laboratories, some can talk knowledgeably about major laboratories of the Air Force, or about the NSF-funded Engineering Research Centers, or about the DOE multiprogram national laboratories. Some claim a comprehensive knowledge at a broader level—university labs, government labs, industrial labs, U.S. labs, Korean labs. The evidence for such claims is modest.

Unfortunately, public policy focuses on the broadest level. In some rare cases, policy makers assume they have a good understanding of, say, industrial labs, and the policies directed toward them, such as tax credits, are allegedly premised on that understanding. More often, public policy is scattershot. Policy makers know they are shooting into the dark but hope to have enough policy pellets to hit something. Policy makers, much like you and we, are ignorant about R&D laboratories. Why do they (and we) know so little? Because there is so much to know. Knowledge of R&D laboratories as a whole is virtually impossible, except at the most superficial level. The result is that there hasn't been much in the way of attempts even to map the system, much less to make sense of it.

Our own mapping efforts (described subsequently in detail) have taken the middle road. It is not possible, probably not even desirable, to compile minute information about every R&D lab. But simply assuming that all university labs are virtually the same, or that weapons labs differ little from each other, or that labs in the food products industry are pretty much the same, impedes the effectiveness of R&D policy.

Incomplete Policy Frameworks

R&D laboratories are the neglected stepchild of public policy. R&D laboratory directors, especially directors of the federal government's national labs, will probably raise their eyebrows at this assertion. Given the number of government policies affecting them, the number of studies conducted by academics, executive agencies, congressional staff, and the U.S. General Accounting Office, it is likely that the federal lab administrators feel poked, prodded, and pummelled. But despite the great attention paid recently to R&D laboratories, as an *institution* they remain neglected.

Public policy for science and technology rarely gives much recognition to the R&D laboratory as a social and political institution. There are implicit "policy frameworks" used to develop strategy, policy, and analysis of science and technology policy issues and three particular policy frameworks dominate. These policy frameworks include: (1) *fields* of science and technology, (2) *human producers* of science and technology, and (3) *missions* and *megaprojects*.

U.S. policy makers routinely entertain policy questions flowing from these three policy frameworks. Thus, the fields policy framework gives rise to the following sorts of questions:

- How can we enhance national security through new defense systems?
- What can be done to accelerate research on superconductivity?
- What is the best strategy for doing basic biotechnology science while at the same time making technology development open to democratic debate?
- How can we identify "critical materials" and support research and development of technology for them?
- What is the best path for developing and commercializing new bio-engineered plants and animals?

These and other such science and technology policy problems focus on the field of science and technology, its needs, and the apparent opportunities presented by the fields.

From the perspective of the human producers policy framework, policy issues revolve around the availability and productivity of categories and types of scientists and technicians. Policy questions of this sort include:

- What can be done to increase high school students' interest in scientific careers?
- Is the U.S. scientific enterprise too reliant on foreign nationals?
- What can be done to increase support for graduate training in physics?
- Is there an overproduction of Ph.D.'s and if so, how is production best managed?

In general, human resources questions flow from the human producers policy framework. The questions within this framework are often easy to articulate even if the answers are not always so straightforward.

The mission policy framework has been a dominant focus for many years. Much scientific debate has been spent on such topics as:

- Should government underwrite technology development for synthetic fuels?
- Should defense R&D focus on dual-use technology?
- Is there a need to invest in R&D for high definition television?
- Should the government conduct research related to scientific and technical standards?
- Who should develop energy options for the long-term future?

- Who should maintain germ plasm for critical plants?
- Should basic physics be international or national in character?

Recently, many of the questions arising from the mission policy framework have been related to megaprojects:

- Who should do what to ensure the development of electric vehicles?
- What are the implications of investing in the R&D required for the Strategic Defense Initiative?
- How will the commitment to a NASA space station affect the R&D from unmanned flyby missions?
- What are the potential economic and social returns from a massive human genome project?

In reality, there is a great deal of overlap in the issues posed by these diverse policy frameworks. Thus, when policy makers were struggling with whether to continue to fund the superconducting supercollider (SSC) in the early 1990s, after hundreds of millions of dollars had already been spent, the decision had implications for the future course of high-energy physics, the need for various categories of technical support personnel, and, of course, the project itself had strong mission implications. Without better knowledge of the role of R&D laboratories, especially federal laboratories, in the U.S. national innovation system, policy makers continue to pour fuel into the innovation engine without regard to the engine's structure, fuel requirements, or on-road performance.

From Incomplete Policy Frameworks to Systemic Analysis

A preeminent concern in our book is viewing U.S. R&D performers as critical elements of the national innovation system. The complexity of U.S. R&D laboratories, the mysteries of the 16,000, and the difficulties of obtaining data about them militate against a systems-level perspective on U.S. R&D laboratories. However, a great many of the problems in U.S. science and technology policy flow from the absence of a system-level perspective.

Policy making for R&D, at least in the United States, has given only limited attention to the characterization of R&D providers as a system. This shortcoming is particularly noticeable as environmental forces acting upon R&D laboratories become more complex and as new organizational forms and new institutional designs emerge. With greater understanding of the U.S. R&D laboratory system, policy making can move beyond traditional stereotypes and give greater consideration to the system-level effects of large-scale policies. For decades, the dominant stereotype for public policy making has been based on sector—government labs, industry labs, and university labs.

There are long-standing stereotypical roles ascribed to the labs of the three respective sectors. For example, universities are seen as the bastion of fundamental research, industry as the home of commercially related, applied, and development research, and while the government lab stereotype is a bit more murky, government labs are often viewed as sites for supporting national research missions, especially in weapons, energy, space, and agriculture.

The policy focus on sectors provides some benefits in addition to its many distortions. The advantages of the threefold sector stereotype are easily discerned. In the first place, there is an obvious validity to that classification. There are universities and they do have labs. There do seem to be differences among laboratories that can be explained by the sector they belong to. Most important, policy makers are accustomed to thinking in terms of the stereotypes, and sector remains one of the few points of distinction readily available to the nonspecialist. In short, the threefold stereotype clearly serves a function. We do not contend that it is no longer reasonable or useful to think in terms of the university-government-industry distinction, only that something more is needed.

The vignettes at the beginning of this chapter suggest the diversity of U.S. R&D labs, a diversity not fully captured in the three sector categories. Many laboratories are hybrids not easily labeled in terms of the distinction. In one study (Crow and Bozeman, 1987a) of more than 200 energy-related R&D labs in the U.S. and Canada, fully one-third did not easily fit into a single sector category. These hybrids include some of the most important laboratories in the U.S. The best known example of a hybrid is the DOE multiprogram "GOCO" labs: government-owned, contractor-operated. But there are many less familiar examples of hybrids, such as the Logistics Management Institute and the Applied Physics Laboratory, the latter of which has over 2,500 employees, supports the Navy, and is run by a largely medical university in Baltimore.

Lab Window 1.3: "What Is It? Fish or Foul?"

Applied Physics Lab, Johns Hopkins University
http://www.jhuapl.edu/

MISSION

The general purpose of The Johns Hopkins University (JHU) can be stated as public service through education, research, and the application of knowledge to human affairs. As part of the university, the Applied Physics Laboratory (APL) shares this purpose through the development and application of science and technology for the enhancement of the security of the USA and for the solution of problems of national and global significance. The laboratory further shares in the university's purpose through basic and applied research and through participation in the educational programs of the academic divisions of the university to which its staff and facilities can make an especially favorable contribution.

Dedicated to public service, objectivity, and problem-solving the Applied Physics Laboratory has maintained an unmatched breadth of capabilities for more than 50 years as a dedicated vital Navy resource and as a university laboratory providing research, development, and engineering support to the Department of Defense to apply effective technical solutions to complex national defense problems. APL conducts collaborative research programs with other JHU divisions and provides an unbiased but expert source of consultation in resolving problems in the development, testing, and production of complex systems. There is no competition with or contracts accepted from industry in areas that fall within APL's central Navy work (to avoid the possibility of conflict of interest).

AREAS OF EXPERTISE

Significant Capabilities
Fleet (theater) defense; combat systems research and development (R&D); guided missile systems R&D; strategic systems evaluations; submarine security and survivability; space R&D; simulations, models, and operations analysis; technical intelligence; related mission R&D.

Systems
Missile, combat, spacecraft, communications, surveillance, radar, command and control, information, transportation.

Technologies
Electronics, information processing, communications, navigation, guidance, propulsion, aerodynamics, oceanography, space physics, sonar, software development, signal processing, materials, biomedicine.

ORGANIZATION
Maintains 140 specialized research and test facilities

STAFFING
The staff of the Applied Physics Laboratory is the basis of our reputation and the key to our success. More than half of APL's staff consists of technical professionals, primarily scientists and engineers; 70% of the technical staff have advanced degrees. APL has a highly versatile staff who understands the complex technologies and compromises that are necessary to obtain practical and balanced approaches to problems. Project managers assemble cross-disciplinary teams and augment the technical capabilities of our personnel for specific projects by using resources from other university divisions or institutions. Of 2,559 full-time staff, there are 785 support staff and 1,774 professional staff. Of the technical staff, 22% are trained in math and computer sciences, 52% in engineering, and 21% in chemistry, physics, and other disciplines.

BUDGET

Estimated funding level approximately $400M through more than 200 separate tasks and various sponsors.

Consider, for example, the Brookhaven National Laboratory on Long Island, New York. The laboratory is owned by the U.S. Department of Energy but is operated by a university and not-for-profit research institute. This operating arrangement makes the laboratory a what? A university center? A government lab? A new category unto itself?

Lab Window 1.4: "Who's Managing the National Laboratory?"

Brookhaven National Laboratory
http://www.bnl.gov

HISTORY

Brookhaven National Laboratory (BNL) is a U.S. Department of Energy (DOE) scientific research laboratory located on Long Island, New York, established in 1947 on the former site of the U.S. Army's Camp Upton. As a nondefense research institution, BNL is dedicated to basic and applied investigation in a multitude of scientific disciplines. The laboratory shares the use of its facilities, many of which are beyond the scope of individual institutions, with scientists from academic, industrial, and government organizations. To provide another instrument for doing forefront physics well into the next century, BNL is now building its next world-class facility: the Relativistic Heavy Ion Collider.

MISSION

Brookhaven's primary objective has always been to gain a deeper understanding of the laws of nature—the necessary foundation for all technical advances. New knowledge is constantly sought in such fields as physics, chemistry, biology, mathematics, medicine, environmental science, and energy technology. Collaborative research and technology transfer with industry are also important parts of Brookhaven's mission.

LOCATION

Long Island

HISTORIC ACCOMPLISHMENTS

Altogether, four Nobel Prizes have gone to physicists who did their research at Brookhaven. One award was for theoretical work; the others were for experiments done at one of the Laboratory's large research machines.

STAFFING
3,200-member staff and over 4,000 visitors who come to BNL every year to use world-class facilities.

OWNERSHIP
DOE-owned. Operated By Associated Universities, Inc., from 1947 to 1997. Founding Universities were: Columbia, Cornell, Harvard University, Johns Hopkins, Massachusetts Institute of Technology, University of Pennsylvania, Princeton, University of Rochester, and Yale.

Associated Universities fired by Secretary of Energy Peña and SUNY Stony Brook was brought in to manage.

Even in those cases where there is no difficulty classifying laboratories by sector, the laboratories within a given category often defy the stereotypical behavior assumed for that category. The R&D universe is populated with many "off-types," such as market-driven government labs like the product development labs of the USDA, university labs eschewing research for product testing or technical assistance (the more than 50 manufacturing technology centers funded by the government and run by universities), and industry labs that are world leaders in fundamental research (IBM, Bell Labs). In some of the largest labs, the R&D portfolio is so broad that no stereotype is likely to prove apt.

Lab Window 1.5: "IBM University"

IBM Watson Research Center
http://www.research.ibm.com/research/intro.html

Research drives several activities within the organizational framework of IBM: (1) accelerating development of mainstream products, technologies and services that are the core of IBM's business; (2) creating new opportunities in emerging fields; (3) broadening the corporation's technical base; and (4) conducting exploratory research across a broad range of areas in science and engineering.

IBM Research was founded at the Watson Scientific Computing Laboratory at Columbia University in 1945. The lab was established by Columbia President Nicholas Murray Butler and IBM President Thomas J. Watson, Sr. Prior to the Watson Lab opening, IBM supported clusters of research and development activities in various sites throughout the company.

IBM Research has generated approximately one-quarter of IBM's 30,000 patents worldwide. IBM researchers have received an array of prestigious international awards over the last 50 years:

- Five researchers have earned the distinguished title "Nobel laureate." In 1986 and 1987 scientists at IBM's Zurich Research Laboratory received the Nobel Prize in physics for the development of the scanning tunneling microscope and the discovery of superconductivity in ceramic materials.

- Four researchers have earned the Turing Award in computer science.
- Thirteen researchers have earned the National Medal of Science and Technology, Buckley Award in condensed matter physics, Wolf Prize in physics, Charles Stark Draper Award and The Institute of Electrical and Electronics Engineers Medal of Honor, both in engineering.

The top 12 most significant accomplishments to come out of IBM research over the years include:

One-device memory cell

Magnetic disk storage

Relational database

Reduced Instruction Set Computer (RISC) architecture

Formula Translation System (FORTRAN)

Scanning Tunneling Microscope (STM)

Thin-film magnetic recording heads

Fractals

High-temperature superconductivity

Token-ring networking

Scalable parallel systems

Speech recognition technology

R&D Policy and the Hazards of Stereotyping: Some Illustrations

The threefold sector stereotype continues to work in some instances. Thus, the stereotypical university laboratory, led by a professor and supported by a small research group made up of graduate students and technicians, remains much in evidence and remains vitally important. But in today's environment the term "university R&D laboratory" encompasses so many and such diverse R&D operations that the term offers as much ambiguity as meaning. University labs are increasingly set within larger structures that have diverse missions, in some cases missions that are not strictly academic.

Consider, for example, the agricultural experiment stations that are operated in each state by the land grant university. These are major research organizations, often involving more than 1,000 staff at a single site. These laboratories are funded by a complicated mix of federal, state, and private sources. Each is a semi-autonomous R&D facility that develops and pursues a research agenda designed to address the research questions that are important to the state government parent and to the agricultural commodity groups of the state. The result is that the political and economic environments of each individual state government influence university laboratories differently. No two of these laboratories are the same.

A good illustration of the uniqueness of a university lab is the Iowa Agricultural Experiment Station operated by Iowa State University. In this research organization there is comparatively limited financing from industry and a strong commitment to the production of public domain knowledge. This focus reflects a local environment of strong government influence and historical commitment to a small to moderate-sized farm production unit. In Hawaii, Michigan, North Carolina, Florida, and other states where agricultural diversity is greater, the production units vary in size and the environment is less dominated by the government and more influenced by market forces, resulting in greater diversity in role and function.

Lab Window 1.6: "Constituency Science"

Iowa Agriculture and Home Economics Experiment Station
http://www.ag.iastate.edu/iaexp/centers.html

These centers serve producers, agribusinesses, rural communities, and policy makers throughout Iowa, across the nation, and around the world. They are administered by the Iowa Agriculture and Home Economics Experiment Station.

Brenton Center for Agricultural Instruction and Technology Transfer
The Brenton Center opened in 1995 as the college's distance education center. It contains high-tech classrooms to link the college with sites across Iowa using the state's fiber optic network for continuing education and extension. Center staff train faculty in the use of new technology to teach courses. A high-tech conference room expands the college's technology transfer capabilities.

Center for Agricultural and Rural Development (CARD)
This center is involved in econometric analysis of the impact of biotechnology and technological change on the financial condition and structure of agriculture. It also focuses on policies related to natural resources and conservation, rural and economic development, trade and agriculture, and food nutrition. The Midwest Agribusiness Trade Research and Information Center and the Food and Agricultural Policy Research Institute are affiliated with CARD.

Iowa Center for Agricultural Safety and Health (I-CASH)
The center was created by state legislation in 1990. It is a partnership of Iowa State University, the University of Iowa, the Iowa Department of Agriculture and Land Stewardship, and the Iowa Department of Public Health and is located at the University of Iowa. Its mission is to coordinate and focus Iowa's public and private resources to establish service, education, and prevention programs to enhance the health and safety of farm families, farm workers, and the rural and agricultural communities.

Iowa Pork Industry Center

Established in July 1994, the center focuses on programs for pork producers and the public that deal with education, field problem solving, demonstration, and consumer issues. Its programs complement and integrate ISU Extension programs.

Leopold Center for Sustainable Agriculture

Named for conservationist Aldo Leopold, this center was established by the Iowa legislature in 1987 to conduct research on the environmental and socioeconomic impacts of farming practices and to help develop profitable farming systems that preserve the productivity and quality of natural resources and the environment.

North Central Regional Plant Introduction Station

One of four regional centers, the station maintains seed (germplasm) collections of 30,000 individual lines. The station has three basic activities: to grow and store seed to maintain viability of the collection, to conduct research, and to serve as a distribution center for plant scientists.

Seed Science Center

Programs in this center include research, seed testing, training seed specialists and seed scientists, and providing information for seed growers, conditioners, and sellers.

Utilization Center for Agricultural Products (UCAP)

Increased utilization of agricultural products through development of new products, new markets, and new processing technology is the focus of the center. It strengthens and broadens programs in two existing ISU centers, the Meat Export Research Center and the Center for Crops Utilization Research. The Food Safety Consortium also is part of UCAP's research efforts.

Consider the University of Florida's Institute for Food and Agricultural Sciences. Field stations located across the state, including major citrus research stations, take on the task of testing for the best insecticide, development of various production and processing techniques, and cooperative breeding experiments with growers. There is substantial federal, state, and private investment in these complex university "labs."

Lab Window 1.7: "Public/Private Cooperation for Orange Juice"

UF/IFAS Citrus Research and Education Center
http://WWW.IFAS.UFL.EDU/www/agator/htm/CREC.htm

The Citrus Research and Education Center (CREC) at Lake Alfred is the largest research, extension, and teaching unit of the University of Florida's Institute of Food and Agricultural Sciences (IFAS). In addition, it includes the scientific research staff of the Florida Department of Citrus (FDOC). CREC is unique among research centers in Florida and the United States in that it focuses entirely on one commodity, making it the foremost institution of its kind. Scientists from various academic disciplines conduct research aimed at increasing the strength and health of the Florida citrus industry and the citrus-consuming public.

The mission of CREC is to meet the developmental needs of the Florida citrus industry through research, extension, and education programs. This is accomplished by: *Solving* the problems facing the industry through research and discovering new fundamental information with which to solve future problems. *Being* an effective transfer of available information at all levels between researchers and industry personnel through a variety of communication methods. *Offering* a series of citrus-related courses at the advanced level for university graduate students and industry personnel.

The staff is composed of approximately 60 scientists from both IFAS and FDOC. The scientists are affiliated with the University of Florida's departments of horticultural sciences, food science and human nutrition, entomology and nematology, plant pathology, soils, food and resource economics, and agricultural engineering. The center's support staff include biologists, chemists, computer technicians, electron microscopists, grove technicians, and clerical business employees. Modern laboratory and support facilities in addition to numerous greenhouses and 250 acres of citrus groves offer an excellent environment for scientific research.

PROGRAM AREAS EMPHASIZED

Plant pathology. The plant pathology perspective encompasses etiology, epidemiology, and disease management strategies for field diseases of both fresh market and processed citrus.

Postharvest. This group's research addresses the needs of the Florida fresh citrus industry with priority given to fruit quality enhancement.

Horticulture. Members provide research leadership in virtually all aspects of citrus biology and production-related practices from seed to harvest.

Entomology/Nematology. Scientists investigate pest management strategies and their integration into management systems for citrus insect, mite, and nematode pests. Biological control receives major emphasis.

Processing. Researchers are engaged in quality preservation, by-product processing technology, microbial control, food packaging, and processing engineering.

Consider the Engineering Research Centers of the National Science Foundation, with their focus on cooperative research with industry. These differ greatly from the university-owned basic astronomy laboratories spread out around the country. The Engineering Design Research Center at Carnegie Mellon University illustrates this point. In the mid 1980s the NSF established this center at CMU in conjunction with several major industries. The purpose of the center is to develop knowledge and techniques for the computer-aided design of manufacturing systems and to transfer these systems to industry. Researchers focus on intermediate-range applied and directed basic research. The research aids in the development of new competitive capacity in American industry and brings about new engineering paradigms in schools of engineering. The ERC agenda is quite different from the work performed at the Kitt Peak National Observatory (operated by a multi-university consortium for NSF) where the focus is on galactic mapping, astrophysics, and related basic research. Interaction with industry is limited. The descriptor "university lab" conveys some meaning but misses many essential nuances.

Lab Window 1.8: "What Sterotype Does This Lab Fit?"

Engineering Design Research Center
http://www.edrc.cmu.edu/

Founded in 1986. The mission of the EDRC is to create, demonstrate, and disseminate cross-disciplinary design methodologies as a significant contribution to the development of design science. The objective of the Synthesis Lab is to develop computational models and computer environments for the generation and selection of alternatives in order to improve the integration of engineering systems at the preliminary stages of design. Emphasis is on the selection of the topology or shape that defines a given system and on the efficient evaluation of alternatives.

A major motivation for synthesis research is that decisions at the preliminary stages of design have a great impact on the cost, quality, and manufacturability of products and processes. Another major motivation is that synthesis, while it lies at the core of design, is still an area that is not well developed or understood. Our work addresses the following questions and challenges:

1. How to create a design space to explore and search among many design alternatives
2. How to develop computer environments that can effectively support design teams and enhance the synthesis process
3. How to generalize computational models for synthesis across various domains
4. How to combine qualitative reasoning and quantitative modeling and how to account for coupling of decisions and complex analysis models in synthesis

The main objective of the Design for Manufacturing Laboratory is to bring downstream concerns (such as manufacturability, assemblability, repairability, and environmental impact) into the earlier phases of the design process. Achieving this objective requires two forms of abstraction. The first abstracts information about the design that is pertinent to the manufacturing process. While the design process produces large volumes of information, only a select subset (such as geometry and properties of the manufacturing medium) is required to support the manufacturing process. The second is abstracting information about the manufacturing process that would constrain the possible design alternatives. These constraints would guide the design process into generating an alternative that could be effectively manufactured with the given facilities and acceptable environmental impact. By moving downstream concerns up into the design process, we should be able to produce higher quality artifacts at lower cost and in less time. The costly experimental cycle of design-prototype-evaluation would be eliminated and replaced by a process that produced a product that was correct the first time. Achieving this objective presents several key issues:

1. How can a methodology be developed that abstracts the essence of each manufacturing process and uses these abstractions to synthesize design alternatives?
2. Can we develop a system capable of reasoning across a range of manufacturing processes and resolving conflicts among constraints imposed by individual manufacturing processes?
3. The effectiveness of an integrated design and manufacturing environment must be demonstrated through the timely construction of artifacts. How do we develop rapid prototyping facilities that embrace a multitude of manufacturing processes?

The center's research is organized into two laboratories addressing the two basic directions of information flow through the product cycle: top-down and bottom-up. The Synthesis Laboratory addresses the early generation and selection of design alternatives, while the Design for Manufacturing Laboratory focuses on the introduction of downstream concerns into the early design stages. The mission of the educational program is to produce a new generation of engineers trained in design methodologies and to disseminate design methods to the engineering curricula and to practicing engineers in the field. The industrial program seeks to improve design practice by developing collaborative efforts between EDRC research in labs, thrusts, and projects and their corresponding receptors across a broad range of industries.

The same diversity can be seen in the industry and government classification. Industrial laboratories can range from the technical service shop of a particular division of a company to a basic research unit heavily funded by government. These differences are acute in that they affect the type of knowledge and technology produced as well as the character of the lab itself. Consider the Bell Laboratories of Lucent Technologies, Inc., formerly of AT&T. These are major research facilities that pursue "directed basic" research aimed at the needs of the Lucent technology complex. Even within a single corporation the laboratories can be very different. The Bell Lab at Murray Hill, New Jersey, with its focus on directed basic research is very different in character and environment from the other lab sites in the Lucent corporate structure emphasizing the development of new communication and data systems for particular markets.

Lab Window 1.9: "AT&T Labs: Several Labs, Not One"

AT&T Labs
http://www.att.com/attlabs/

ADVANCED TECHNOLOGIES LABORATORY studies, enhances, and supports diverse levels of communication operations through software development, performance analysis, and operational research. The lab also facilitates development of worldwide standards for applications, networking, and software technology.

ADVANCED COMMUNICATIONS LABORATORY focuses on integrated wired/wireless-based access technologies involving cellular, private mobile radio, personal communications services, and satellite systems.

NEW CONCEPTS LABORATORY develops service concepts for applications within the AT&T Network. By working with both the technical and business community within the corporation, as well as by monitoring the telecommunications industry, the technical staff seeks to understand future customer trends in broadband and multimedia services in residential and commercial markets.

SOFTWARE PRACTICES AND TECHNOLOGY LABORATORY partners with service development organizations to transform software engineering breakthroughs into sustainable practices. Drawing on the best from AT&T Labs research and the software community at large, the lab applies independent assessment techniques and software engineering technology as part of AT&T's continuous improvement program.

TECHNOLOGY REALIZATION LABORATORY supports the development and introduction of new communication services from technology conception and market validation through deployment, spanning the entire technology realization process.

Consider the differences between the McDonnell Douglas Corporate Research in St. Louis (an early 1990s series of changes in the lab and the pending 1997 merger of McDonnell Douglas with Boeing have essentially eliminated this lab's purpose) and the now defunct Technical Services Center of Northern Petrochem in Illinois. In the McDonnell Douglas case the corporate research was almost entirely funded by government for the purpose of conducting long-range applied research. The results of the laboratory's work generally had the character of a public good available in the open literature. The laboratory originated in the 1950s at the urging of the Department of Defense (DOD). The DOD was interested in the development of basic knowledge within its prime contractors as a means of helping them to stay competitive and current with scientific advances. The researchers in this setting operated very similarly to university researchers. At the Northern Petrochem Lab the focus was technical service to the clients and to the operating divisions. The operation of the firm's complicated chemical plants required constant research support as a part of normal operations. The research function supported the various clients of the parent corporation in the use of products or processes. The results of this research process were propriety in character. The entire function and purpose of these two laboratories was completely different.

Government laboratories, likewise, can be focused on the production of public knowledge, or they can be designed to produce knowledge for the consumption of a single firm, sector, or industry. For instance, the Commonwealth of Kentucky is operating a laboratory that focuses on the production of knowledge and technology for consumption by the local mining and coal utilization industry with little contribution to scientific literature. In Ft. Collins, Colorado, the U.S. Department of Agriculture (USDA) operates a seed germ plasm research lab that maintains seed samples of most of the important varieties of selected crops. This seed laboratory collects, analyzes, and stores for general distribution all of these samples as a public good. It is intended for the benefit of all potential users. The term "government lab" can cover vast differences.

Lab Window 1.10: "Seeds of the Future"

National Seed Storage Lab, USDA
http://www.ars-grin.gov:80/ars/NoPlains/FtCollins/nsslmain.html

The National Seed Storage Laboratory (NSSL) is a core facility for the U.S. National Plant Germplasm System. The NSSL is located on the Colorado State University Campus in Fort Collins, Colorado. The primary mission of the laboratory is to provide long-term preservation of plant genetic resources of important plant species for current and future generations. There are two operational groups at the NSSL: preservation and research.

Throughout the world plant genetic diversity is being eroded. Loss of these plants will result in the loss of genes for disease and insect resistance, high yield, improved nutrition, new pharmaceuticals, and crops that can overcome global climate changes.

The National Seed Storage Laboratory (NSSL), as a member of the National Plant Germplasm System, preserves the genetic diversity of crop plants from all over the world (more than 248,000 collections). Because many commercial crops grown in the U.S. are not native, these plant introductions have been vital for developing our food and fiber crops. Plant breeders have used these introductions and other source materials to develop high-yielding varieties with disease and insect resistance. This has enabled farmers to increase yields and lower costs so that the average U.S. family now spends less than 12% of its income for food. When varieties have genetic resistance to disease and insects, reduced use of pesticides helps provide a cleaner environment.

The mission of the Seed Viability and Storage Research Unit, National Seed Storage Laboratory, is to effectively document, preserve, and maintain viable seed and propagules of diverse plant germ plasm in long-term storage, to develop and evaluate procedures for determining seed quality of accessions, and to provide administrative support to allow for effective operation of this unit. The mission also includes the distribution of seed, when not available from the active collections, for crop improvement throughout the world.

Specific objectives are

1. To preserve the base collection of plant germplasm for the National Plant Germplasm System
2. To determine initial quality and to periodically monitor viability of plant germplasm placed into storage

3. To continuously maintain and update the National Seed Storage Laboratory database, the Germplasm Resources Information Network (GRIN)

4. Improve cooperation and coordination among germplasm curators to identify priorities for long-term germplasm preservation

5. Develop methods to preserve plant propagules of species and accessions not currently in the base collection

6. Develop and improve technologies for evaluating viability, vigor, genetic integrity, and potential longevity of preserved germplasm

7. Evaluate conventional and cryogenic storage protocols and develop strategies to improve cost efficiency

8. Conduct pilot studies to evaluate protocols for long-term preservation of plant propagules in order to transfer technology to germplasm curators, seed companies, and other customers

---

The National Comparative Research and Development Project: Constructing Empirically Based Policy Paradigms

Our ideas about R&D laboratories and their roles in national innovation systems have been shaped by our experiences in designing and carrying out the National Comparative Research and Development Project (NCRDP), one of the most comprehensive efforts to develop an empirically based profile of national innovation systems (NIS). The National Comparative Research and Development Project began in 1984 as an attempt to understand the workings of the NIS in the United States and other industrial nations. More than 40 researchers in four countries have been directly involved in the NCRDP, developing data, writing research reports, and helping to design new aspects of the project. Over the course of the past thirteen years, we have interviewed or received questionnaires from thousands of scientists, science administrators, and science policy makers, chiefly in the U.S. but also in Canada, Japan, Korea, Germany, Taiwan, and England.

The U.S. component of the NCRDP was formulated to address the two related problems facing anyone interested in more systematic, better rationalized policy for R&D laboratories: the role of R&D laboratories in the NIS system and the deployment of U.S. federal laboratories in the service of commercial goals and "competitiveness." Along the way, the NCRDP has resulted in a wide variety of substantive findings about the NIS in the United States and also in a set of general propositions and policy design assumptions.

Throughout the book we draw from the data and empirical findings of the NCRDP. In the interests of coherence, we do not provide a great deal of technical information about the database of the NCRDP (which is actually an umbrella name for five different and relatively large databases). Appendix 1 provides a summary of the NCRDP and its various stages and gives some insight into the course and temporal flow of the project. Greater detail can be found in the various papers and publications associated with the NCRDP.

THE NEXUS OF DATA AND PHILOSOPHY

The studies provided in the NCRDP are not, for the most part, manifestly prescriptive. But in this book we make no bones about our prescriptive intentions. Our policy design objectives proceed in two directions, from the data up and from "policy paradigms" down.

In our view, the fundamental premises (or policy paradigms) for science and technology policy are too confining, not wrong but incomplete, even immature. We begin by briefly outlining the dominant assumptions or policy paradigms for science and technology policy. We proceed to develop an alternative, which we will call "institutional design."

It should not be surprising that with over 14,000 new R&D labs having been designed and built since 1950 and with the growth of the national R&D budget from under $1 billion in 1940 to more than $200 billion in 1997, the need for institutionally relevant R&D policies is acute. R&D institutions and R&D policies have grown without consideration of system-level issues, coordination issues, life-cycle issues, or even consistent review and evaluation. All of these facts, matched with our data, indicate that policy makers are ill prepared to deal with what is now a 50-year-old enterprise in need of serious renovation.

When policy makers employ policy paradigms but have little or no empirical base, they take unnecessary and avoidable risks. Perhaps worse, they may not know when or to what degree they have succeeded or failed. Thus, system-level data for the NIS goes hand in hand with the development of new ways of thinking about U.S. science and technology policy.

SCIENCE AND TECHNOLOGY POLICY PARADIGMS:
MARKET, MISSION, AND COOPERATIVE

From our nation's earliest days, leaders have wrestled with the difficult question of how far government should go in regulating or influencing individual and commercial behavior. On the one hand, we seek a government that is not intrusive—a government that does not meddle unduly in private affairs. The *market failure paradigm* has historically been the cornerstone for U.S. science and technology policy. As a performer of R&D, the chief orientation for a government role lies within the *mission paradigm* that government should perform R&D in service of well-specified missions in which there is a national interest not easily served by private R&D. The most important element of the mission policy philosophy is defense and national security-related R&D, but such missions as energy production and conservation, medicine and public health, space, and agriculture have expanded the role of federal laboratories. There is also widespread recognition of the unique ability of government to marshal resources and to influence events in such a way as to foster technology development and innovation.

According to the *cooperative science and technology paradigm*, government's role can be that of a broker—performing research and developing technology to be consumed by industry. Federal laboratories are at the center of this policy philosophy as they are the institutions providing the scientific and technical resources upon which the cooperative technology paradigm is built. The cooperative science and technology paradigm has always been part of the nation's science and technology policy, the oldest example being the system of agricultural extension stations, first established in the 1880s at land grant colleges started in the 1860s and 1870s. But recently interest in the cooperative technology paradigm has been renewed, spurring political controversy. In the next section, we review briefly the premises of these three R&D policy philosophies and, in a later section, we suggest and outline a fourth alternative.

The Market Failure Paradigm

The market failure paradigm for science and technology policy (and its attendant economic development implications) is based on familiar premises: that free markets are the most efficient allocators of goods and services and, left to its own devices, an unfettered market will lead to optimal rates of science production, technical change, and economic growth. The market failure approach recognizes that there may be a role for government in science and technology policy when there are clear externalities (i.e., that benefits cannot be captured in the market), when transaction costs are extremely high, and when information is unavailable or there are distortions in information so that market signals are not clear. Most U.S. public policy, not only laboratory policy and not just science and technology policy, is strongly influenced by the market failure paradigm.

Despite recent experiments with an expanded government role in science- and technology-based economic development, the market failure paradigm in science and technology policy is alive and well. For example, such comparatively recent initiatives as the R&D tax credit provisions from the Economic Recovery Tax Act of 1981 (recently extended) assume industry is the progenitor of innovation and the role of government is chiefly limited to staying out of the way (Bucy, 1985; Bozeman and Link, 1984; Bozeman and Link, 1983; Eisner, 1985; Landau and Hanney, 1981). Similarly, efforts to roll back early initiatives of the Clinton administration, such as NIST's Advanced

Technology Program and the Manufacturing Extension Partnerships Technology Applications Policy, are framed within the assumptions and rhetoric of the market failure paradigm.

In early 1997 the Republican majority of the 105th Congress issued a declaration to end "corporate welfare." Using the market failure argument, the politicians targeted for cuts a number of federal programs designed to stimulate new technology development in critical industries, including all programs at NIST and DOD directly linked to civilian technology development.

In its simplest form, the market failure paradigm assumes that there is a significant need for new scientific and technical resources for economic productivity and that the private sector will, through the competitive workings of the market, sense that need and respond in an economically efficient manner. Thus, the market paradigm assumes little or no role for federal laboratories as contributors to the NIS, beyond the role fulfilled by their public domain mission functions.

Historically, the market failure paradigm has been dominant in science and technology policy excepting defense or space-related R&D and R&D directly relevant to agency missions. Until the early 1980s there was a widespread consensus that technological innovation and commercialization were the preserve of industry. Until the 1980s there was much validity to the stereotype that government R&D was for missions, industry R&D for technology innovation and development, and universities for basic knowledge and people production.

The Cooperative Technology Paradigm

During the period of economic uncertainty in the 1980s and the perceived crisis in U.S. competitiveness many core assumptions were reexamined, including bedrock faith in the private sector as the source of virtually all significant innovation (M.I.T. Commission on Industrial Productivity, 1989; National Academy of Sciences, 1978; National Governors' Association, 1987; President's Commission on Industrial Competitiveness, 1985; Council on Competitiveness, 1993; Niosi, 1988). As other nations, especially Japan, forged ahead with government support of technology development, the market failure paradigm seemed less compelling (Bozeman and Crow, 1990; Crow and Bozeman, 1987b).

In the 1980s a number of policy initiatives challenged the preeminence of the market failure paradigm for technology policy with a new one, a cooperative technology paradigm. As used here, the cooperative technology paradigm is an umbrella term for a set of values emphasizing cooperation among sectors—industry, government, and university—and cooperation among rival firms in the development of precompetitive technologies and "infratechnologies" (Link and Tassey, 1987). Kash and Rycroft have provided strong arguments for cooperative technology policy approaches (Kash and Rycroft, 1994; Kash, 1989; Rycroft and Kash, 1994).

The 1980s and early 1990s were a period in which the dominant market failure paradigm received its strongest challenge. Challenges to market failure thinking included policies changing patent policy to expand the use of government technology (Patent and Trademark Laws Amendment, 1980), relaxing antitrust, promoting cooperative R&D (Link and Bauer, 1989; National Cooperative Research Act of 1984), establishing research consortia and multisector centers (Smilor and Gibson, 1991), and altering the guidelines for the disposition of government-owned intellectual property (Bagur and Guissinger, 1987; Gillespie, 1988). Finally, the Clinton/Gore technology plan of then presidential and vice-presidential candidates Bill Clinton and Al Gore (Clinton and Gore, 1992) outlined a national plan for a movement toward enhanced cooperation, national science and technology planning, and national technology projects.

The cooperative technology development policies having attracted the most attention are those pertaining to domestic technology transfer, especially the use of federal laboratories as a partner in the commercialization of technology (Herrmann, 1983; Rahm, Bozeman, and Crow, 1988; U.S. General Accounting Office, 1989). Federal laboratories had previously been aloof from commercial concerns, indeed had been prohibited by law from developing technology specifically for private vendors, but the legislation of the 1980s gradually changed their mission, tenor, and climate (Bozeman, 1994b) and, to some extent, the companies interacting with the labs (Roessner and Bean, 1991). The intellectual property dictum "if it belongs to everyone, it belongs to no one" began to take hold as the government labs increasingly moved from a sole focus on public domain research to a mandated role as a technology development partner to industry.

Recently the pendulum seems to have swung away from the cooperative technology paradigm. While there are still many elements of the policy model evident in federal laboratory policies and activities, the enthusiasm of the first two years of the Clinton administration barely survived the election of the 1994 and 1996 Republican-majority Congresses. The Clinton administration now presses its cooperative technology paradigm more gingerly. The Republican Congress continues to oppose an expanded role for federal laboratories.

The Mission Paradigm

The earliest government involvement in science and technology policy was within the framework of the mission paradigm. The mission paradigm assumes that the federal laboratories' role in science and technology should flow directly from the duly authorized missions of agencies and should not extend beyond those missions in pursuit of more generalized technology development, innovation, or competitiveness goals. As such, this view is not radically different from the market failure paradigm, except that the mission paradigm encourages the performance of nondefense missions in a wide array of policy domains.

The roots of the mission paradigm can be traced to early government R&D roles in national defense, public health, and, to some extent, agriculture (which also has had cooperative technology paradigm elements). But recently there have been several reminders of the significance of mission paradigm thinking. The Secretary of Energy Advisory Board's report (1994) on the future of the Department of Energy national laboratories, widely known as the Galvin Commission report, emphasized that many labs had strayed from their traditional energy- and defense-related missions and that any future work in technology development and commercialization should flow directly from those missions. In many ways, this was a sharp reaction against policy initiatives and program expansion under the cooperative technology paradigm. At the same time, the counterreaction to "dual technology" initiatives has much the same tone—that the defense mission comes first and that efforts too far afield may well undermine the defense mission.

Table 1.1 provides a summary of the assumptions of each of the respective policy philosophies.

THREE PROBLEMS IN FEDERAL R&D POLICY MAKING

While the policy paradigms outlined above continue to provide what is in many ways a sensible framework for science and technology policy making, they do not adequately address some of the continuing problems in policy making for federal laboratories. Before offering an alternative policy paradigm, "institutional design," we discuss briefly some limitations of existing frameworks.

Problem: Poor Base of Empirical Knowledge

The U.S. R&D laboratory system is remarkably complex, with thousands of extremely diverse labs influenced by a congeries of rapidly changing environmental forces, many of which are poorly understood. Gathering useful information is a costly and otherwise formidable task. Likewise, developing richer conceptualization and categories for laboratories is challenging. Where does one begin? Labs vary in so many significant dimensions—technology, organization, resources, sector, culture, market orientation—there is no obvious point of analytical departure.

The U.S. faces a dilemma. The national laboratory system may be viewed as an aging, increasingly feeble knowledge infrastructure. But despite widespread concern that the current configuration of R&D assets leaves much to be desired, the would-be policy doctors know very little about the physiology

| Table 1.1 | Three Competing R&D Policy Models |

	Market Failure	Mission	Cooperative
Core Assumptions	1. Markets are most efficient allocator of information and technology. 2. Government role limited to market failures such as extensive externalities; high transaction costs; and information distortions. Small mission domain, chiefly in defense. 3. Innovation flows from and to private sector, minimal government role.	1. The government role should be closely tied to authorized programmatic missions of agencies. 2. Government R&D is limited to missions of agencies, but not confined to defense. 3. Government should not compete with private sector in innovation and technology. But a government role in connection with traditional activities of line agencies.	1. Markets not always the most efficient route to innovation and economic growth. 2. Global economy requires more centralized planning and broader support for civilian technology development. 3. Government can play a role in developing technology, especially precompetitive technology, for use in the private sector.
Peak Influence	Highly influential during all periods.	1945–1965; 1992-present.	1992–1994.
Policy Examples	Deregulation; contraction of government role; R&D tax credits; capital gains tax roll back. Little or no need for federal laboratories except in defense support.	Creation of energy policy R&D, agricultural labs, and other such broad mission frameworks.	Expansion of federal laboratory roles in technology transfer and cooperative research; manufacturing extension policies.
Theoretical Roots	Neoclassical economics.	Traditional liberal governance with broad definition of government role.	"the 'new' industrial organization theory."

of their patient and, as a result, are as likely to kill as to cure. Many feel something must be done to renew the U.S. R&D laboratory system. But only the rashest actors entertain *re*-configuring without a knowledge of the current configuration.

Problem: R&D Policy and the Hazards of Stereotyping

To a large extent, policy making is by stereotype: university labs produce basic research, government labs serve agency missions, industry labs develop and commercialize technology. This threefold stereotype continues to work in some instances. But in today's environment the term "federal laboratory" encompasses so many and such diverse R&D operations that the term offers as much ambiguity as meaning. The same is true of industry and university labs.

In the absence of greater empirical knowledge of the structure, environments, and performance of laboratories within the U.S. R&D laboratory system, stereotypes provide a ready shorthand. But in most instances laboratory stereotypes distort more than they edify. The "pure" university lab focusing on basic research yields to the entrepreneurial university lab charged with an economic development mission; the industry lab, serving its parent and immune from government influence, takes its place alongside private-sector government contractors more attuned to government R&D agendas than to any company's needs; the classic postwar government weapons lab finds itself refashioned as environmental savior, promising strides in hazardous waste R&D. All too often, stereotypes simply get in the way of effective policy making.

Problem: Too Much Ideology and Not Enough Pragmatism

U.S. science and technology policy and the federal laboratory system revolve to a surprising degree around ideological premises. This is not to say that ideology supersedes politics, but that politics is cloaked in ideology. Among discussions of such big issues as the power of the market, the potential of activist government, and the meaning of globalization, the more mundane issues of "what works?" easily get lost. Often the dominance of ideology in science and technology debates is less due to hard and fast positions of disputants than to lack of evidence. If one has little evidence of "what works," then ideology at least provides some decision-making ballast. One key to more pragmatic approaches to policy making—and perhaps less dramatic politically based swings of the policy pendulum—is the availability of additional evidence about the federal laboratory system. Another is change in the policy-making apparatus for the labs. In recent times the federal laboratory system has been caught between the Scylla of bureaucratic micromanagement and the Charybdis of congressional macromanagement. The chief "stakeholders" of the laboratories, industry users and the scientific community, influence policy making only episodically through participation on ad hoc panels, laboratory advisory panels, and congressional testimony. With a Congress steeped in political necessities and a bureaucracy serving multiple masters, these stakeholders may be the best constituency for a policy based on "what works."

An Alternative Science and Technology Policy Paradigm: The Empirically Based Pragmatism of Institutional Design

Let us consider an alternative that complements the three paradigms: the *institutional design paradigm*. The key point of the institutional design paradigm for science and technology policy is to view the R&D laboratories as the important contributor to the NIS and to take a pragmatic view in assessing contributions. The findings from the NCRDP suggest the utility of an institutional design approach. Whereas most policy deliberations give great attention to determining "the appropriate government role," our studies show that R&D labs have for many years played a wide variety of roles despite continued efforts to put each sector in a defined niche.

The institutional design paradigm can be viewed in terms of just a few policy-making "principles." These principles are discussed below (and summarized in Table 1.2).

THE SYSTEMIC PRINCIPLE

Public policy rarely reflects a systemic view of R&D laboratories, but some public policy analysts have noted the lack of systems thinking and the need to think about the health of national innovation systems. The fundamental idea of a systemic perspective is simple enough, namely, that a nation's role in the increasingly interconnected global economy depends upon its effectiveness in developing and exploiting innovation resources provided by its significant producers of scientific knowledge, technology, and technical know-how. A first premise, then, and one shared in most analyses of science and technology policy, is that innovation has a major effect on the rate and direction of economic growth, standards of living, quality of life, and international competitiveness (Nelson, Peck, and Kalachek, 1967; Nelson, 1962; 1984). Second, it is not only the *amount* of science and technology resources available to a nation that determines the strength of its NIS but also the configuration of those resources, their distribution, and the ability of the nation to exploit the resources of its NIS. Third, like most social systems, the NIS is in a state of constant flux. A NIS with a high degree of environmental complexity and interdependence, such as that of the U.S., is particularly dynamic, and it is particularly difficult to track its changes. A highly complex NIS is generally quite sensitive to a variety of societal and market changes including not only science and technology polices but also other policies that indirectly affect the NIS, such as tax policy, antitrust policy, defense policy, and indeed most categories of public policy. The NIS is affected not only by public policies but by a wide variety of social and economic trends, ranging from the savings rate, interest rates, and balance of trade to such factors as availability of skilled labor and general population dynamics such as birth trends and patterns of immigration.

Given the interdependence of the NIS, its complexity, its changeability, and the many environmental forces affecting it, tracking change in the NIS presents a challenge. But a lack of systemic knowledge undercuts policy objectives.

The Comparative Advantage Principle

Different labs do different things well. Public policies should consider R&D laboratories' comparative advantages. This finding occurs again and again in NCRDP research (e.g., Bozeman and Crow, 1988; Crow and Bozeman, 1991). R&D laboratories are involved in a tremendous variety of technical activities involving a wide range of outputs in all substantive fields of science and engineering. Even the largest laboratories specialize, and no laboratory does everything well. Public policies often seem to ignore comparative advantages. Questions center on such general issues as whether to dismantle government labs or on possible shortfalls in industrial R&D. But in R&D policy the best answers are rarely simple ones. For example, it may make sense to dismantle some government labs and to expand others. There may be a shortfall in industrial R&D in some fields and a surfeit in others. The comparative advantage assumption reinforces a systemic view in that it underscores the fact that different actors play different roles within the NIS.

The Player Principle

Of the more than 16,000 U.S. R&D laboratories, only a small fraction has the capacity to contribute on a sustained basis to either scientific breakthrough or technological innovation (Bozeman and Crow, 1988). As many as 65% of all U.S. laboratories are small, engineering job shops providing important services to the firm, but services that rarely have much external impact. The market failure paradigm assumes correctly that these "non-player" laboratories are generally better off being left out of government technology policy deliberations. This benign neglect of the masses provides some hope of accumulating valid and reliable policy-relevant information about the "players," especially the 500 or so high-capacity laboratories responsible for most sweeping science and technology change. The "players" should be understood, monitored closely, and targeted for most interventionist R&D policies. The majority of labs, left to the forces of the market, remains as a foundation of the economy, while the players fuel tomorrow's economy.

The Never-Neutral Principle

As private laboratories come to be more and more subject to political and public policy influence, their character begins to change, sometimes fundamentally (Crow and Bozeman, 1987a). It is possible to graft a technology transfer function on a weapons lab or a basic research center. It is possible to redirect a large-scale technology development lab into a technical assistance lab. Sometimes these added functions are met with considerable success. But invariably the introduction of new functions affects old ones. Sometimes new functions complement old ones, and often they compete. As policy makers seek to alter federal laboratories, they do well to consider the net impact of change on core functions.

Consider the Ames Laboratory of the U.S. Department of Energy. For more than 40 years (1942–1988) the principal mission of the lab was basic materials science. In the ensuing 10 years, the lab's mission expanded to include technology transfer, environmental waste management research, and limited technology development. The result has been a dramatic drop-off in the conduct of the basic chemistry, physics, and math that underpin materials sciences and a transformation of the core mission and purpose of the lab. Thus, the lab today is a part of a larger technology institute with a science- and technology-based economic development mission.

Lab Window 1.11: "Expanding Missions for a Tiny Laboratory"

Ames Laboratory, DOE, IA

http://www.ameslab.gov/

MISSION

Ames Laboratory effectively focuses diverse fundamental and applied research strengths upon issues of national concern, cultivates tomorrow's research talent, and develops and transfers technologies to improve industrial competitiveness and enhance U.S. economic security. At the forefront of current materials research, high performance computing, and environmental science and management efforts, the laboratory seeks solutions to energy-related problems through the exploration of physics, chemistry, engineering, applied mathematics, and materials sciences. All operations are conducted so as to maintain the health and safety of all workers and with a genuine concern for the environment.

Ames Lab now pursues much broader priorities than the materials research that has given the lab international credibility. Examples of specific projects include: world-class fundamental photosynthesis studies to ultimately help in the design of synthetic molecules for direct solar energy conversion; development of a remote-controlled analysis system that will acquire and analyze samples from hazardous waste sites at greatly reduced risk and cost; research to break free of traditional programming methods and harness the power of the most advanced computing systems available for scientists to unlock the secrets of revolutionary new materials like superconductors, fullerenes and quasicrystals; and the synthesis and study of nontraditional materials such as organic polymers and organometallic materials to serve as novel semiconductors, processable preceramics, and nonlinear optical systems.

Through Ames Laboratory, the U.S. Department of Energy

1. Provides approximately 30% of all research dollars coming into Iowa State University

2. Provides advanced materials to industry, university, and government research communities through its internationally recognized Materials Preparation Center. Industrial users include AT&T, Pratt & Whitney, and Rhone-Poulenc.

3. Operates a Materials Referral System and Hotline, which handles more than 800 materials inquiries annually from companies worldwide

4. Contributes technical expertise to Iowa companies such as Sauer-Sundstrand, Burlington Northern Railroad, and Pioneer Hi-Bred International through the Iowa Companies Assistance Program

STAFFING
Approximately 650 employees, including more than 300 scientists and engineers

BUDGET
Estimated annual budget of $35 million

OWNERSHIP
Government-owned, contractor-operated research laboratory of the U.S. Department of Energy (DOE). Iowa State University operates Ames Laboratory for the DOE.

The Opportunity Cost Principle

The implications of the "never neutral principle" for *policy* are reflected in the "opportunity cost principle" for *policy evaluation*. Laboratory evaluations often include a marginal cost-benefit element. But especially during a time of flat or declining resources, it is not enough to know whether an activity has a positive net benefit. Might that same set of resources lead to a much higher net benefit if used some other way? For example, if DOD laboratories redirect their efforts to "dual-use technology," there is no reason to believe that the core mission of defense support will remain unaffected. The question, then, is not simply whether there is potential for dual-use technology but what the opportunity costs are. Since there are opportunity costs associated with new missions (Bozeman and Fellows, 1988), evaluating R&D laboratories and policies requires an evaluation model that explicitly recognizes opportunity costs (Bozeman and Coker, 1992).

Why Focus on Federal Laboratories?
The Leverage Points for the NIS

Consistency may, indeed, be the hobgoblin of little minds. While we are not prepared to say that inconsistency is the hallmark of larger minds, we are prepared to *appear* inconsistent. On the one hand, a major argument of this book (see especially chapter 4) is that knowledge of a lab's sector or ownership is insufficient. On the other hand, much of the book is devoted to federal laboratories. The reason for this is simple: if one wishes to redesign the R&D laboratory system, the federal laboratories provide a leverage point.

Table 1.2	An "Institutional Design" Paradigm for Science and Technology Policy

→ **"Player Principle":** policy and strategy focus is on those labs, probably 200–300, with significant real potential to contribute to public domain science, innovation, and national economic development.

→ **"Systemic Principle":** policies and labs evaluated empirically and holistically, in terms of impact on NIS and labs' niche within the NIS

→ **"Never Neutral Principle":** new functions (e.g., technology transfer) often have dramatic effects on existing functions (e.g., basic research).

→ **"Comparative Advantage Principle":** public policy should be differentiated and targeted; based on labs' mix of resources and capabilities (and not ownership), laboratories should be reinforced to do what they do well.

→ **"Opportunity Cost Principle:** requires that any new missions be judged in terms of net impact compared to alternative uses of those resources.

The prerequisites of institutional design:

1. Greater knowledge of laboratory assets, capabilities, and performance;
2. Greater coordination and creation of coordinating apparatus;
3. Reduced role for line agency management;
4. Strategic management flexibility, decentralization, and autonomy
5. Greater policy-making role for external stakeholders and users

Policy makers exert direct control over federal laboratories. Not that industrial labs are exempt from government influence. But the gentle nudge that policy makers give industrial labs by altering the tax treatment of R&D or by providing subsidies does not compare to the ability to create (or terminate) labs and missions.

Another reason to focus on federal laboratories is that this is an era of great reflection (and even some action) about the role and purpose of federal labs. It is fair to say that federal laboratories are in a state of siege and transformation. There is a great deal of flux and uncertainty. The outcome of this transformation process will say much about what we want to be as a nation in the next century, as these labs represent our collective interests and desires.

FEDERAL LABORATORIES STATE OF TRANSITION

Evidence of the trials of federal laboratories requires no deep search through government archives, only a reading of the daily newspaper. Sometimes federal laboratories are simply in the line of fire as critics take aim at the line agencies of which they are a part. So long as influential members of Congress remain set on closing down the Departments of Energy and Commerce, the federal laboratory system is not safe (Goodman, Brownlee, and Watson, 1995; Browning, 1995). In other instances, the heat is on the laboratories themselves. The weapons laboratories in particular have drawn close scrutiny (Smith, 1993; Markusen et al., 1995).

Many threats to federal laboratories arise from large-scale social and political changes over which laboratories have no control. But sometimes the laboratories and agencies governing them seem to be their own worst enemies. Consider some of the following events and allegations:

- DOD labs relearn in 1996 that chemical weapons did contaminate US troops in 1991 in Iraq, contrary to earlier claims by DOD scientists (Radetsky, 1997).
- Federal laboratories have been accused of waste due to poor accounting procedures (Hanson, 1992).
- The Department of Energy Inspector General's office accused DOE and a federal lab of mismanaging cleanup of contaminated land (Hedges, 1990).
- A whistle-blower calling attention to the vulnerability of a government nuclear research plant was punished (Lippman, 1989).
- Strangest of all, ghosts from the 1950s came back to haunt as insidious nuclear experiments on humans finally came to light.

The adage that "all publicity is good publicity" does not pertain to public bureaucracies. When such items are taken into consideration along with the allegations by the former Office of Technology Assessment about the inability of federal labs to implement defense conversion (Smith, 1993; *Business Week*, 1993), General Accounting Office concerns about the disrepair and maintenance problems of federal labs, and widespread suspicion that the federal laboratories' new cooperative technology and technology transfer efforts are lacking (Saddler, 1992; Lepkowski, 1991), one gains perspective on the labs' state of siege. Perhaps the crowning blow: among product reviews of leaf blowers and home fitness equipment, a *Consumer's Digest* article on "The War on Washington Waste" complained (erroneously) that the "Energy Department spends one-fifth of its budget on cooperative energy-development programs, giving money to firms like General Electric and Westinghouse to support research, which they then turn to their profit" (Hager, 1996).

THE STAKES

In our view, the fate of the U.S. federal laboratories is a matter of great consequence. Whether or not one agrees with the view that the federal laboratories are "a reservoir of scientific and technological talent that can help America to compete in international markets" (Schriesheim, 1990), and whether or not one is impressed with the 58 Nobel laureates in the national laboratory system (U.S. Department of Energy, 1994), the resources devoted to federal laboratories cannot help but command attention. More than $23 billion per year are spent on R&D in 726 federal labs (Crease, 1993), more than one-third of all federal R&D money expended. Federal laboratories employ more than 60,000 scientists and engineers, a significant fraction of the U.S. scientific and technical human capital. In addition to producing tens of thousands of scientific and technical publications and papers each year, federal laboratory personnel file nearly a thousand patent applications each year (U.S. Department of Energy, 1994d). The range of functions performed by federal labs is remarkable. The core functions of such "superlabs" as Sandia National Laboratory and the Naval Research Lab are familiar, but the superlabs have a wide range of functions that go well beyond their core technical mission. Indeed, this mission accretion has been the source of considerable criticism; White House Science Council, 1983). But if the largest labs receive the lion's share of attention, 700 or so less visible federal labs undertake an even more diverse array of scientific and technical tasks, ranging from collecting and analyzing seed samples (the National Seed Storage Laboratory of the USDA, Ft. Collins, Colorado) to devising building materials that will resist terrorist attacks (the Army Construction Engineering Research Laboratory, Urbana, Illinois). Federal laboratories are engaged in research and development activities at all points of the spectrum: basic, precommercial applied, direct applied, development, and testing. They are involved not only in R&D but provide extensive technical assistance and permit access to technical and equipment resources that are in many instances unique. The question "What should we do with the federal laboratories?" goes well beyond issues of cost savings and managerial efficiency.

In sum, the federal labs' state of transformation actually represents an opportunity. When the old assumptions begin to be questioned, sometimes momentous good things happen. Sometimes momentous terrible things happen. Perhaps the most likely among the "momentous terrible thing" scenarios is the gutting of the federal laboratory system that has evolved during the past 50 years and includes some of the most significant scientific resources in the world. It is not difficult to imagine this scenario. Positive outcomes require more imagination. We hope this book presents some ideas that can contribute to the dialog about just what a "positive outcome" for the federal laboratory system (and, more generally, the NIS) is and maybe even some ideas about how to achieve those outcomes.

If we wish to understand the role of R&D laboratories in the NIS, a better understanding of the term "national innovation system" should be part of that process. In the next chapter we examine the concept, its origins, and applications and consider the relevance of R&D laboratories to the NIS.

CHAPTER TWO

Laboratory-Based Innovation in the American National Innovation System

Chris Tucker and Bhaven Sampat

Systematic designs for R&D laboratories almost always focus on particular R&D organizations rather than on sets of laboratories. *Organization design* is commonplace, whereas *institutional design* is rare indeed. Such efforts as the Packard Report (White House Science Council, 1983) or the Galvin Commission Report (U.S. Department of Energy, 1994a) represent a groping toward a design of sets of laboratories. But in our view, "design" seems too strong a word for the loosely integrated sets of recommendations provided by these and other blue-ribbon panels.

Organizational design occurs daily within the U.S. R&D laboratory system. Industrial labs decide to link their central R&D facility to their production unit, university labs consolidate labs in biology with those in chemistry, government labs change the boxes and lines of programmatic organization charts. R&D managers and policy makers *think* in terms of organizations and organizational design. In part, this is due to control. People in authority often find organizational change within their grasp; institutional design requires much more cooperation and negotiation. In part, the reluctance to engage in institutional design stems from a long-standing distrust of planning and "social engineering." But one reason for limited design activity relates more to the concept than to practicalities. Managers and policy makers have little history of thinking in terms of institutions, and there are few models to follow. It simply does not occur to many policy makers to think of R&D laboratories as collectivities playing special roles in a national innovation system (NIS). If the director of a small, engineering-oriented laboratory answering only to a corporate parent fails to think in terms of institutional design there is no reason for alarm or surprise. If individuals engaged in making national science and technology policies fail to think in terms of institutional designs, the ability to achieve national science and technology goals is impaired.

The absence of a *concept* for institutional design is the most fundamental way policy is limited by design. A first step is simply to develop a concept of the system within which institutional design occurs. In this chapter we present concepts of the national innovation system and review the intellectual bases of work focused on the NIS.

What Is a "National Innovation System"?

The term "national innovation system" refers to the complex network of agents, policies, and institutions supporting the process of technical advance in an economy. As we emphasize below, different authors have (implicitly or explicitly) included different things under the rubric of the NIS. Following Lundvall (1992), we distinguish between use of the term NIS in the "narrow" sense and in a "broad" sense. Scholars looking at the "narrow" NIS focus on organizations and institutions directly involved in the processes of scientific and technological searching and exploring. The "broad" NIS concept includes all economic, political, and other social institutions affecting learning, searching, and exploring activities. Thus the narrow definition would include a nation's patent system, its universities, and its research laboratories, while the broad definition also includes many other subsystems and processes, such as a nation's financial system, its monetary policies, norms of competition, and the internal organization of private firms in the nation.

From the preceding paragraph, it should be clear that the concept of the national innovation system is a flexible one, and (as we discuss below) different aspects are stressed by different authors depending both on their theoretical perspectives and on the particular country being studied. However, the essence of the concept is the focus on nation-specific factors affecting innovation. Underlying the concept of the "national" innovation system is the argument that there are important differences at the national level in policies and institutions affecting the rate, direction, and character of technical advance in a nation's firms and industries.

Though the concept of the national innovation system was first explicitly used in the mid-1980s, most scholars agree that the idea goes back at least to the German economist Friedrich List. This nineteenth-century German economist is perhaps best remembered for his advocacy of infant industry protection to fuel industrialization. List actually advocated a broad range of policies to promote technological learning, innovation, and economic growth; policies that he hoped would help Germany (and other "less developed" countries) catch up economically with England (Freeman, 1995).

While a review of the NIS literature is beyond the scope of the present study, the interested reader is advised and encouraged to refer to what are commonly regarded as the two key contributions to the literature, *National*

Innovation Systems: A Comparative Analysis, edited by Richard Nelson, and *National Systems of Innovation: Towards a Theory of Innovation and Interactive Learning,* edited by Bengt-Ake Lundvall.

Varying Sources and Patterns of Innovation

The NIS is an institutional lens of analysis, finding technological innovation within a country to be conditioned by the institutions underpinning different sectors. While some of these institutions are sector specific, others are more general, affecting innovative activity across a range of sectors. This lens of analysis, taken in a broader form, may similarly be considered institutional as it takes seriously varieties of industrial structures, corporate forms, and business strategies as institutional forms in their own right.

Knowledge bases of innovation vary dramatically by sector (from automobiles and biotechnology to construction and optics/imaging) and this variance suggests different industrial dynamics and institutional needs. The differences between similar sectors of different countries (e.g., steel production in the U.S. and Japan) should also be understood to have their roots in core sets of technologies, industrial structures, and supporting institutions within a particular national environment.

Below is an attempt at a short history of the evolution of the U.S. national innovation system that attempts a broad account of the rise of laboratories as the chief source of innovative activity across many key sectors in American society. We hope it demonstrates some of the diversity in lab structures, missions, capabilities, publicness, and relationships.

Sources and Patterns in the American System of Manufactures

Any history of the evolution of the current American innovation system is best served by first understanding the mode of innovation of the mid-nineteenth century, dubbed the "American System of Manufactures" (Rosenberg, 1969). This system was characterized by a mass-market orientation that required rapid, volume production of inexpensive, quality goods; experience with large-scale plant design; unit operations in production and their integration into coordinated flow; the building of cheap machines rather than the mere use of machines; and standardization and interchangeable parts. The goal was both to reduce the costly role of labor in production and to continually develop production processes for the manufacture of novel end products requiring more sophisticated production techniques (Trescott, 1981; Hounshell, 1984; Rosenberg, 1963).

Sources of innovation in the "American System" were decidedly craft-based and were fundamentally interwoven with the evolution of the machine-tool industry (Rosenberg, 1963). Many products were assembled from machined components, whether as parts for more complex machines or as commercial products. Industrial labs, almost exclusively chemical labs, were founded for the purpose of testing, grading, and purifying substances used in other productive activities (Mowery and Rosenberg, 1989). But from these humble beginnings sprung the corporate R&D laboratory and the innovation capability that ultimately evolved into the world's foremost NIS.

RISE OF UNIVERSITY SCIENTIFIC RESEARCH AND ENGINEERING

The emergence of industrial lab work was made possible by the development of industrially pertinent scientific knowledge and the training of scientists capable of serving industry. In the U.S. in the early nineteenth century, science was largely not appreciated for what it might offer in support of industry. The beginning of the nineteenth century saw the creation of science professorships at U.S. colleges and the founding of scientific societies and their journals—with little attention paid to the potential commercial importance of science (Bernal, 1953). America continued to lag in the generation of advances in fundamental knowledge about natural phenomena (Veysey, 1965). The sort of public support enjoyed by the French system of *écoles*, German state universities, and the British support of various research and education institutions was nearly absent in the American experience prior to the 1870s (Dupree, 1957). With the exception of West Point and some early state-funded colleges, the bulk of the institutional weight supporting American scientific work depended on private philanthropy and religious organizations—and was few and far between.

The institutional landscape supporting scientific and technical advances in the early nineteenth century in the United States did not support cutting-edge scientific advances. It is also clear from the calls for reform that these institutions did little to support American technical and industrial advances in the early and mid-nineteenth century (Rudolph, 1962). The early application of scientific knowledge to technical advance in American chemical work, for instance, was supported largely by émigrés and Americans returning with degrees from abroad (Haber, 1958; Ihde, 1964). As various U.S. industries came to depend on particular bodies of scientific knowledge for the advancement of their technical base, there was naturally a concomitant rise in the dependence on institutions supporting research and education.

In part, this inadequate institutional suite was due to the failure of the scientifically concerned to overcome the controlling disposition of the American colleges (Rudolph, 1962). Scientific advance was not one of the organizing principles of the early colleges. Nevertheless, as early as the 1820s the dissatisfaction of some colleges led to the introduction of scientific studies and modern languages as a replacement for ancient languages and religious studies (Geiger, 1986; Veysey, 1965). This was emulated by other institutions in their development of parallel course study (that is, classical and scientific) (Guralnick, 1979). However, this trend was not adopted by many other institutions (Rudolph, 1962). Nor did scientific knowledge in the U.S. context develop to the point of widespread industrial application.

Still, academe was being adapted to new goals and to new social conditions. Reformers who introduced a new institutional logic of "social utility" to a few of these schools were driven, in large part, by the desire to bring the institutional capacities of the academic institutions to bear on the practical problems of conquering the wilderness of the continent and erecting centers of commerce (Rudolph, 1962).

This new logic included applying science to practical problems, an element that came to the fore around the middle of the century in the creation of scientific schools at enlightened institutions such as Harvard, Yale, Columbia, Dartmouth, and Clark (Veysey, 1965; Nelson and Rosenberg, 1994). These institutions had their forerunner in the Rensselaer Institute, founded in 1824, which anticipated the rationale of the American land-grant college (Ricketts, 1934). Its purpose was to train teachers who could go out into the district schools and instruct "the sons and daughters of farmers and mechanics . . . in the application of experimental chemistry, philosophy, and natural history, to agriculture, domestic economy, the arts and manufactures," such that useful knowledge—at least some of it scientific—be applied to the business of living (Ricketts, 1934). When reorganized in 1849 according to the progress made in technical education by French scientific schools, Rensselaer began to place more emphasis on fundamental laboratory research in chemistry and physics and reworked its civil engineering program—a model that was widely adopted and that pointed American colleges in a new direction.

Just prior to this reorganization (1846 and 1847), Yale and Harvard had been institutionalizing scientific research and education in a way then unheard of in their peer institutions. With philanthropic grants these institutions created new professorships dedicated to the sciences, mostly applied, and began scientific degree programs, initially for undergraduates. This activity led to the creation of Yale's Sheffield Scientific School and Harvard's Lawrence School. Though held in lower esteem than traditional programs, the idea of the scientific school in the 1850s spread rapidly to many other colleges and universities (Guralnick, 1979). These institutions brought science even closer to direct application in worldly activities. But their relationship to industrial activity was weak and the reliance of industrial processes on science was nearly nonexistent.

Perhaps the culmination of the institutionalization of the utility principle (linking science to practical needs) in higher education was the creation of the land-grant college in the Morrill Act of 1862 (Eddy, 1957; Ross, 1942).

This act placed federal moneys in the hands of states and empowered them to set aside land grants on which they could build universities. This mechanism led to an entirely new network of institutions that was dedicated to the advancement of the agricultural and mechanical arts (Dupree, 1957). It took much of the 1860s and the early part of the 1870s to implement the land-grant scheme across the states. This created the institutional girth necessary to shift the entire U.S. system of higher education, both publicly and privately supported, toward an organizing principle of utility. Popularly inspired, the land-grant undertaking expanded those areas of practical life that were considered possible foci for scientific and technical advance. Rather than excluding the sciences, the land-grant concept called for the support of practical skills without diminishing the role of fundamental research, the humanities, and the advancement of culture (Veysey, 1965). Not only did these institutions educate largely rural populations in technical and scientific skills, they also served as the institutional base for research extension work, as the Hatch Act of 1887 established in each state experiment stations that benefited a wide range of agricultural sectors (Rosenberg, 1977).

A very different set of motivations led to the institutionalization of scientific research in the U.S. With the development of a network of scientists who became increasingly conscious of their national community and its relative standing on the global scene, there was increased activism for the establishment of academic institutions that were dedicated to the support of research into fundamental scientific phenomena (Daniels, 1971). The German model of graduate education and its basic research orientation were offered as benchmarks by which the state of American academic institutions should be measured (Veysey, 1965).

This activism called forth its patrons, and in 1876 Johns Hopkins was established with the exclusive mission of supporting fundamental research and graduate education (Geiger, 1986). Hiring the best researchers from U.S. and foreign universities, Hopkins set the new standard among the elite universities in the U.S. The establishment of the University of Chicago in 1890 raised the count of American universities begun for the support of *fundamental* scientific research to two. After this, other elite institutions undertook reorganization to better accommodate the emerging specialization of the sciences, applied sciences, and engineering. Places such as Columbia University reorganized in recognition of the reality of the new independence of these scientific disciplines (Geiger, 1986; Veysey, 1965). Laboratory investigations now had a secure place in academic institutions, a place where new scientific, applied science, and engineering disciplines could flourish.

"One of the major accomplishments of the U.S. universities during the first half of the twentieth century was to effect the institutionalization of the new engineering and applied science disciplines" (Rosenberg and Nelson,

1994:327). Between the 1890s and the first decades of the twentieth century, applied fields such as chemical engineering, electrical engineering, and aeronautical engineering became established in U.S. universities—disciplines that both reflected and solidified new kinds of close connections between American universities and a variety of U.S. industries. "The rise of these new disciplines and training programs in American universities was induced by and made possible the growing use of university-trained engineers and scientists in industry, and in particular the rise of the industrial research laboratory in the chemical industry and the new electrical equipment industries, and later throughout industry" (Rosenberg and Nelson, 1994:327).

It was the growth of these disciplines and their corresponding technical communities that allowed for linkages across universities and also across universities and industry. These relationships were systematic and cumulative, providing a better flow of knowledge between fundamental laboratory pursuits and applied scientific work of all types (Rosenberg and Nelson, 1994).

RISE OF LABS AS A SOURCE OF INNOVATION IN THE AMERICAN EXPERIENCE
In the U.S. laboratory activity related to industrial needs first found its place in processes of materials testing, grading, and standardization. This occurred in support of different industrial sectors beginning in the 1830s and became more institutionalized throughout the rest of the nineteenth century on a sector-by-sector basis depending on the industry's needs and the ability of the sciences. The first instances of the commercial impact of science has been characterized as

> not only—and perhaps not even primarily—as the emergence of new bodies of scientific knowledge that were subsequently applied to industry. Rather, a rapidly growing and industrializing economy encountered all sorts of situations in newly emerging technologies where further improvement and progress required drawing on the existing fund of scientific knowledge. – *Mowery and Rosenberg, 1989*

Since the fundamental scientific knowledge employed during this period was already in existence, there was no significant research element actively seeking to push back the fields' frontiers. Instead, science in the industrial context was "engaged in a variety of routine and elementary tasks such as the grading and testing of materials, assaying, quality control, and writing of specifications" (Mowery and Rosenberg, 1989).

This application of science often was institutionalized in the creation of market-driven independent labs that took on research by contract. They conducted research on questions farmed out by industry and advised businessmen on their operations. But this type of activity was neither systematic nor widespread. Independent research evolved where there was a market.

Many industrial labs developed around questions of chemistry and materials. Traditional chemical concerns, steel producers, and extractive industries were the most direct beneficiaries from early laboratory science (Mowery and Rosenberg, 1989). Some of these industries represented the first instances of industrial research being integrated into a larger production unit. But it was the emergence of America's science-based industries in the last part of the nineteenth century that saw the evolution of the laboratory structure that is considered the origin of the modern industrial R&D lab. These labs altered the character of the industrial research enterprise, introducing organized industrial research laboratories aimed at the creation of science-based technological systems or products, as opposed to merely deploying scientific know-how for the testing of elements of craft practice. The commercialization of electrification by Edison, Brush, Thompson, and Westinghouse and communication applications by Bell required the harnessing of fundamental natural phenomena for the creation of new technologies (McMahon, 1984; Hughes, 1983; Wasserman, 1985).

Lab Window 2.1: "From Bell Labs to Lucent Technologies"

http://www.lucent.com/news/factbook/factbook.html

The heritage of Western Electric and Bell Labs is at the core of Lucent Technologies.

1869   Elisha Gray and Enos Barton form a small manufacturing firm in Cleveland, Ohio.

1872   Gray and Barton's firm renamed Western Electric Manufacturing Company.

1876   Alexander Graham Bell files patent application for the telephone.

1881   American Bell purchases controlling interest in Western Electric and makes it the manufacturer of equipment for the Bell Telephone companies.

1899   AT&T, created in 1885, takes over American Bell Telephone and becomes parent to Western Electric and the Bell System companies.

1925   Bell Laboratories is created from the AT&T and Western Electric engineering departments, which had been combined in 1907.

1937   Dr. Clinton J. Davisson wins the Nobel Prize for his experimental confirmation of the wave nature of electrons. He becomes the first of seven Bell Laboratories scientists to win the Nobel Prize.

1946   Western Electric manufactures a then record 4 million telephones.

1947   Bell Labs invents first transistor in Murray Hill, New Jersey.

1956   AT&T signs consent decree limiting Western Electric to manufacturing equipment for the Bell System and the U.S. government.

1979   A Federal Communications Commission inquiry restricts AT&T from selling enhanced services except through an AT&T subsidiary, American Bell, which begins operations in 1983.

1982   AT&T and the Justice Department settle an antitrust suit that modifies the 1956 consent decree. AT&T will spin off local telephone companies and retain long-distance service, Western Electric, and Bell Labs.

1984   AT&T spins off local telephone companies. Western Electric's charter is assumed by a new unit called AT&T Technologies.

1995   AT&T announces its plan for restructuring into three separate, publicly traded companies: a services company that will retain the name AT&T; a systems and technology company (Lucent Technologies) composed of Bell Labs, systems for network operators, business communications systems, consumer products, and microelectronics; and a computer company, which returned to the NCR name.

1996   Henry B. Schacht, a member of AT&T's board of directors since 1981, is named chairman and chief executive officer of Lucent Technologies, formerly AT&T's systems and technology company. Lucent Technologies is launched with an initial public stock offering in April and completes its spin-off from AT&T on September 30.

These systemic technologies required research for the initial design, diversification, and rationalization of systems and created the need for better fundamental understanding of associated phenomena to ensure the success of the technological enterprises. Thus these industrial R&D labs helped underpin a range of business strategies, helping to open up technological opportunities these firms could then exploit while also developing technological capabilities making them more able to respond flexibly to challenges. They also allowed these firms to identify new technologies and new firms to purchase in order to better their competitive standing.

As table 2.1 suggests, there was tremendous growth in the number of industrial research labs formed in the late nineteenth and early twentieth centuries. Although many of the pioneers of these science-based industries were not formally educated inventors, much of their informal education depended on the decades of scientific activity undertaken in the universities and supported by the university-centered scientific and technical communities. Many of these pioneers in science-based industries relied upon formally educated scientists in the everyday activities of their labs. As these industries developed, the crucial role played by universities in supporting applied sciences,

Table 2.1	**The Formation of Industrial Research Labs, 1890–1930**				
1890	4	**1905**	100	**1920**	519
1895	22	**1910**	180	**1925**	728
1900	49	**1915**	310	**1930**	1030

Source: Thackray, Sturchio, Carroll, and Bud (1985)

such as chemical engineering, electrical engineering, and so forth should not be underestimated. These industries became increasingly embedded in and reliant upon scientific activities in the universities.

Publicly funded laboratories came to be chartered offering support for technological innovation across a range of sectors (Dupree, 1957). The cooperative agricultural extension laboratories that were built on the land-grant colleges were the first notable example of public authority supporting a system of laboratories. Publicly funded, these labs brought scientific and technological know-how to the agricultural sectors, supporting both producers and processors. From the grading and testing of soils and fertilizers to the analysis of production methods and machinery—not to mention the scientific and technical support of the complex enterprise of food processing—these labs offered the technological and scientific expertise necessary for the various agricultural sectors to enhance their productivity and the quality and diversity of their products. The U.S. Geological Survey, begun in 1869, provided similar scientific services to extractive industries and those industries dependent upon natural materials.

Lab Window 2.2: "Earth Science in the Public Service"

U.S. Geological Survey
http://www.usgs.gov

As the nation's largest earth science research and information agency, the USGS maintains a long tradition of providing "Earth Science in the Public Service."

The USGS, a bureau of the U.S. Department of the Interior, was established to provide a permanent federal agency to conduct the systematic and scientific "classification of the public lands and examination of the geological structure, mineral resources, and products of the national domain."

The mission of the USGS is to provide geologic, topographic, and hydrologic information that contributes to the wise management of the nation's natural resources and that promotes the health, safety, and well-being of the people. This information consists of maps, databases, and descriptions and analyses of the water, energy, and mineral resources, land surface, underlying geologic structure, natural hazards, and dynamic processes of the earth.

As a nation we face serious questions concerning our global environment. How can we ensure an adequate supply of critical water, energy, and mineral resources in the future? In what ways are we irreversibly altering our natural environment when we use these resources? How has the global environment changed over geologic time, and what can the past tell us about the future? Will we have adequate supplies of quality water available for national needs? How can we predict, prevent, and mitigate the effects of natural hazards?

The USGS collects and analyzes the scientific information needed to answer these questions, and disseminates these results to the public in many forms, such as reports, maps, and data bases.

Another governmental research function can be found in the evolution of its role in the establishment and maintenance of standards. Developed first by the Coastal Survey in the early 1800s, the government's standards function eventually evolved well past its original interest in weights and measures (Dupree, 1957). The establishment of the National Bureau of Standards in 1901 brought with it the consolidation of several governmental activities focused on helping develop industry-wide standards for the testing and purification of substances. Some of these laboratory activities grew out of federal commitments to extractive industries implicit in the work of the U.S. Geological Survey and mining-related agencies. Others grew from the pressures for professionalization and efficiency within industry and government posed by the Progressive movement of the late nineteenth and early twentieth century (Wiebe, 1967).

The years around the turn of the century saw the broader establishment of industrial research in science-based industry. Edison's operations, renamed General Electric, and Bell established the industrial R&D labs often held up as paragons of industry research organizations (Hammond, 1941; Wise, 1985; Carlson, 1991; Wasserman, 1985). DuPont established its labs in support of new business directions, recognizing the vulnerability of its black powder operations (Hounshell and Smith, 1988). The electrochemical concerns of the Niagara region, industries such as Union Carbide, ALCOA, Eastman Kodak, and Dow Chemical, maintained labs in support of both process and product innovations (Boundy and Amos, 1990; Graham and Pruitt, 1990; Collins, 1990). Labs came to support the innovation processes of critical industries of the economy, producing both end products and intermediate inputs for sectors not considered science-based, such as automobiles, aircraft, machine tools, textiles, and agriculture.

The first two decades of the twentieth century saw a rapid rise of the science-based industries. University responses to technical human resources needs were monumental, underpinning the laboratory work in each industry. At the same time, universities worked to define their own research agendas.

Wartime Demonstration of the Utility of Organized R&D

World War I was the proving ground for organized industrial research labs, and university-industry-government cooperation in R&D (Dupree, 1957; Mowery and Rosenberg, 1993). High profile successes of organized industrial research, such as processes for the fixation of atmospheric nitrogen and those underpinning petrochemical advances, convinced American industry that investment in organized R&D would bring concrete results in new products and more efficient plant operations.

This realization brought with it an explosion in industrial R&D throughout the 1920s (Mowery, 1981). Concentration in the chemical, electrical, and communications industries continued, but the growth of petrochemical interests and the rapid diffusion of formal R&D into non-science-based industries added to the boom (Mowery and Rosenberg, 1993). Between 1921 and 1927, employment of scientists and engineers in industrial research in U.S. manufacturing firms leapt from 2,775 to 6,320. A considerable amount of the boom in industrial R&D during this period can be attributed to the boom in the transportation equipment and related industries as automobiles came to utilize more complex and sophisticated inputs in the common categories of glass, steels, rubbers, paints, and machined parts and as the government came to support R&D in the aircraft industry (Mowery and Rosenberg, 1993).

This expansion was almost exclusively development-oriented industrial R&D. In recognition of the lack of industry-funded basic research, then Secretary of Commerce Herbert Hoover consistently worked to encourage industry funding of basic research (Dupree, 1957). Explaining the interest industrial concerns had in supporting the advance of fundamental knowledge, he argued for a minor role for government in cooperation with industry.

Grand Transformation

The depression did not bring with it a significant reduction in industry's investment in R&D nor did the New Deal bring significant increases in federal support for R&D (Mowery and Rosenberg, 1993). Universities experiencing financial troubles undertook a number of organizational responses to cope with scarcity (Geiger, 1986). Clearly, the watershed event for U.S. research and development was World War II and the resulting Cold War.

World War II: Organizing Research for War

Prior to the Second World War, the lines between industrial research, government research, and university research were reasonably clear. There was a well-established pattern of university consulting for industry and industrial support of university research. Similarly, government research had its public functions, but they were often cast as the support of industries whose character was ill-suited to the provision of their own infrastructure and technological needs. World War II saw the mobilization of many *private* resources for *public* purposes with public funds (Stewart, 1948; Baxter, 1946; Thiesmeyer and Burchard, 1947; Brooks, 1996). Universities, which had traditionally funded their research through tuition revenues and philanthropy, suddenly saw their faculties contracted and mobilized by the military for the creation of fundamental scientific work and technological solutions to military problems (Geiger, 1986; Geiger, 1992). Corporate R&D capabilities were mobilized for similar purposes. New labs were created to attack scientific and technological problems associated with military missions and to support the training of those required to develop and command these new technologies. In support of the war effort the Office of Scientific Research and Development (OSRD) organized virtually all R&D activity in the nation, civilian and military. The level of funding, the size of the projects, and the type of the applications were entirely new. And just as with World War I, the fruits of federally funded R&D were duly noted.

VANNEVAR BUSH AND THE ENDLESS FRONTIER

World War II drove home the realization that science and the labs in which it was conducted had immeasurable effects on human existence. Security, health, and economic welfare could all be served by the work of scientific and technical personnel in laboratory activity. The director of the wartime Office of Scientific Research and Development (OSRD), Vannevar Bush, issued a report, *Science: The Endless Frontier*, to President Roosevelt arguing just this point and calling for a system of peacetime federal support of scientific research (Bush, 1945). He called for support of health research, something that had already been embraced politically. He assumed that the federal government would continue to support the military missions with mission-oriented R&D, realizing that this would require massive investment in both applied and basic research. At the center of his recommendations was a call for the creation of a National Research Foundation that would, under the guidance of the scientific elite and through the financial support of the federal government, support basic science.

Bush stated that only five American universities (apparently California, Cal Tech, Columbia, Harvard, and Yale) had adequate research conditions in their science departments and that beyond these five, there existed an acute need for supplemental research resources in order to let the staff devote more than half their time to research (Bush, 1945; Geiger, 1986). It was one of his central themes that universities should be at the center of this growth effort. One major result of the political bargain struck in response to *Science: The Endless Frontier* and rival plans was the growth of American research universities as performers of federally funded research and the dispersion of serious research talent from the few to the many. With this growth came a dispersion of funding (Geiger, 1992; Graham and Diamond, 1997). However, as table 2.2 suggests, much of the funding was concentrated in the "elite" universities.

Table 2.2	Concentration of Funding: Federal R&D Obligations to Top Ten Universities (%)				
1952	43.4	**1968**	27.7	**1988**	21.5
1958	37.0	**1978**	23.0		

Source: Geiger (1992)

While Congress fought over the form postwar federal support of science and technology was to take, the Office of Naval Research took the lead in supporting university science and engineering. This ONR support maintained the basic science base while funding military technology projects (Sapolsky, 1979, 1990; Office of Naval Research, 1987). Defense-oriented research of all kinds was later funded by the military in universities and industry. The National Science Foundation was created in 1950, though underfunded, demonstrating the meager political agreement about science for science's sake (England, 1983). Nonetheless, the federal government clearly took over the role of academic patron and expanded academic science to new heights.

In support of the defense missions, corporate research labs came to be funded at a rate that sometimes exceeded 60% of their total funding. Aircraft, communications, electronics, nuclear power—many industries deemed critical to national security experienced significant support for their R&D activities (Mowery and Rosenberg, 1993; Brooks, 1996; Nelson, 1984).

The postwar climate also offered the rationale for the continued support and expansion of the national weapons labs that were developed as part of the Manhattan project. Due to fear of military control of nuclear technology, the Atomic Energy Commission (AEC) was created to develop nuclear power for civilian uses and to govern the use of radioactive material for all purposes, including military (Orlans, 1967; Hewlett and Anderson, 1962). It supported scientific investigations into this technology, built the national labs to focus on nuclear energy and weapons, and regulated the use of nuclear technology. With control over the knowledge base required for the development of nuclear weapons, the AEC also provided the military with lab access

Table 2.3   Federal Support of Academic R&D in Billions and as Percentage of Total

Year	Total academic R&D (M$)	Federally supported R&D (M$)	Federal percentage of total
1935	50	12	24
1960	646	405	63
1965	1,474	1,073	73
1970	2,335	1,647	71
1975	3,409	2,288	67
1980	6,077	4,104	68
1985	9,686	6,056	63
1990 (est.)	16,000	9,250	58

SOURCES: 1935 Nat. Resources Council (1938); NSF (1991)

and services. The AEC maintained and expanded many of the weapons labs that had been created during World War II, resulting in nine multiprogram laboratories: Argonne, Brookhaven, Idaho National Engineering Lab, Oak Ridge, Sandia, Lawrence Livermore, Lawrence Berkeley, Los Alamos, and Pacific Northwest, each among the largest scientific facilities in the world. These labs employed thousands of scientists across many disciplines, focusing chiefly on designing, testing, and manufacturing nuclear weapons systems components and investigating a wide range of basic sciences pertaining to nuclear warfare operations. The AEC, along with the military, funded 76% of academic research in 1950—the proportion rises to 87% if Federal Contract Research Centers are counted (Geiger, 1993). This seemed a productive marriage of civilian know-how and military need.

The astounding growth of the National Institutes of Health (NIH) began before Vannevar Bush even spoke of an "endless frontier." NIH funded both intramural and extramural research, largely at universities. This funding was in support of basic scientific investigations but had wider impact. The growth in extramural funding was a significant part of the expansion of academic medical centers. Basic science investigations led disciplinary development and research specialization. NIH continued to fund this diversification and specialization in labs and research agendas. Basic science advances served to enhance instruction in medical schools and the specialization of medical knowledge. In addition, the federal government funded the construction of hospitals and medical schools through the 1946 Hill-Burton Act (Graham and Diamond, 1997). The hospital aspects of academic medical centers served, as they still do, as laboratories of a sort, providing an arena for continued clinical research and experimentation with new medical procedures. These institutions served as the infrastructure upon which a range of industries have relied, including pharmaceuticals, medical devices, and later the biotechnology industries (Orsenigo, 1989; Rosenberg, Gelijns, and Dawkins, 1995).

Lab Window 2.3: "What Impact Has the NIH Had on the Health of the Nation?"

National Institutes of Health
http://www.nih.gov/

NIH research played a major role in making possible the following achievements of the last few decades:

- Mortality from heart disease, the number one killer in the United States, dropped by 41% between 1971 and 1991.
- Death rates from stroke decreased by 59% during the same period.

- Improved treatments and detection methods increased the relative five-year survival rate for people with cancer to 52%. At present, the survival gain over the rate that existed in the 1960s represents more than 80,000 additional cancer survivors each year.
- Paralysis from spinal cord injury is significantly reduced by rapid treatment with high doses of a steroid. Treatment given within the first eight hours after injury increases recovery in severely injured patients who have lost sensation or mobility below the point of injury.
- Long-term treatment with anticlotting medicines cuts stroke risk from a common heart condition known as atrial fibrillation by 80%.
- In schizophrenia, where suicide is always a potential danger, new medications reduce troublesome symptoms such as delusions and hallucinations in 80% of patients.
- Chances for survival increased for infants with respiratory distress syndrome, an immaturity of the lungs, due to development of a substance to prevent the lungs from collapsing. In general, life expectancy for a baby born today is almost three decades longer than it was for one born at the beginning of the century.
- Those suffering from depression now look forward to returning to work and leisure activities, thanks to treatments that give them an 80% chance to resume a full life in a matter of weeks.
- Vaccines protect against infectious diseases that once killed and disabled millions of children and adults.
- Dental sealants have proved 100% effective in protecting the chewing surfaces of children's molars and premolars where most cavities occur.

Molecular genetics and genomics research has revolutionized biomedical science. In the 1980s and 1990s researchers performed the first trial of gene therapy in humans and are able to locate, identify, and describe the function of many of the genes in the human genome. Scientists predict this new knowledge will lead to genetic tests to diagnose diseases, such as colon, breast, and other cancers, as well as to the eventual development of preventive drug treatments for individuals in families known to be at risk. The ultimate goal is to develop screening tools and gene therapies for the general population, not only for cancer but also for many other diseases.

The development of laboratories capable of serving diverse missions allowed both the achievement of public policy goals and the emergence of many science-based industries that now are part of everyday life. Without the public support of the institutions that advance the pertinent sciences, the emergence of entire productive sectors would have—at the very least—been severely retarded. Just as World War I made industry aware of the benefits of organized R&D efforts, World War II drove home the point to the American public and its leaders that there was a legitimate role for government in the support of science and technology.

Continuing Expansion of R&D for Commercial Use

The massive funding of research in universities brought with it graduate education and technology transfer through trained scientists and technicians. Many industries benefited from the diffusion of this knowledge, bringing core capabilities that allowed them to undertake their own distinctive innovation strategies. Other industries benefited from government contracts for research procurements (Danhof, 1968). The development of technologies for the fulfillment of these contracts led to the evolution of comparative advantages in related downstream commercial technologies. The 1950s and 1960s witnessed a further expansion in the use of organized laboratory R&D in many areas, including nuclear energy, petrochemicals, pharmaceuticals, electrification, communications, and chemicals. The increased dependence of U.S. industrial society on science and R&D laboratories characterized this period and the ones following.

Sputnik as the Second Impetus

The launching of the Soviet satellite Sputnik in 1957 focused U.S. attention on the relative state of our defense technology base (Geiger, 1993). With this, two organizations were created to remedy our lagging position. The National Aeronautics and Space Administration (NASA) was put together from the old National Advisory Council on Aeronautics (NACA) construct that had served (1915–1958) to support American aircraft development for military and then civilian uses (Bilstein, 1989).

Lab Window 2.4: "Before Sputnik: 45 Years of Industry-University-Government Aeronautical Research"

National Advisory Committee for Aeronautics (NACA)
http://www.hq.nasa.gov/office/pao/History/SP-4406/cover.html (Bilstein, 1989)

"In 1915, congressional legislation created an Advisory Committee for Aeronautics. The prefix 'National' soon became customary, was officially adopted, and the familiar acronym NACA emerged as a widely recognized term among the aeronautics community in America. . . .

"The genesis of what came to be known as the National Advisory Committee for Aeronautics (NACA) occurred at a time of accelerating cultural and technological change. Only the year before, Robert Goddard had begun experiments in rocketry and the Panama Canal had opened. Amidst the gathering whirlwind of the First World War, social change and technological transformation persisted. During 1915, the NACA's first year, Albert Einstein postulated his general theory of relativity and Margaret Sanger was jailed as the author of *Family Limitation*, the first popular book on birth control. Frederick Winslow Taylor, father of 'Scientific Management,' died, while disciples like Henry Ford were applying his ideas in the process of achieving prodi-gies of production. Ford produced his one millionth automobile the same year. In 1915, Alexan-der Graham Bell made the first transcontinental call, from New York to San Francisco, with his trusted colleague, Dr. Thomas A. Watson, on the other end of the line. Motion pictures began to reshape American entertainment habits, and New Orleans jazz began to make its indelible imprint on American music. At Sheepshead Bay, New York, a new speed record for automobiles was set, at 102.6 mph, a figure that many fliers of the era would have been happy to match. . . .

"American flying not only lagged behind automotive progress, but also lagged behind Euro-pean aviation. This was particularly galling to many aviation enthusiasts in the United States, the home of the Wright brothers. True, Orville and Wilbur Wright benefited from the work of European pioneers like Otto Lilienthal in Germany and Percy Pilcher in Great Britain. In Amer-ica, the Wrights had corresponded with the well-known engineer and aviation enthusiast, Octave Chanute, and they had knowledge of the work of Samuel P. Langley, aviation pioneer and secretary of the Smithsonian Institution. But the Wrights made the first powered, controlled flight in an airplane on 17 December 1903, on a lonely stretch of beach near Kitty Hawk, North Carolina. Ironically, this feat was widely ignored or misinterpreted by the American press for many years, until 1908, when Orville made trial flights for the War Department and Wilbur's flights overseas enthralled Europe. Impressed by the Wrights, the Europeans nonetheless had already begun a rapid development of aviation, and their growing record of achievements under-scored the lack of organized research in the United States. . . .

"Sentiment for some sort of center of aeronautical research had been building for several years. At the inaugural meeting of the American Aeronautical Society, in 1911, some of its members discussed a national laboratory with federal patronage. The Smithsonian Institution seemed a likely prospect, based on its prestige and the legacy of Samuel Pierpont Langley's dusty equipment, resting where it had been abandoned in his lab behind the Smithsonian 'cas-tle' on the Mall. But the American Aeronautical Society's dreams were frustrated by continued infighting among other organizations which were beginning to see aviation as a promising research frontier, including universities like the Massachusetts Institute of Technology, as well as government agencies like the U.S. Navy and the National Bureau of Standards. . . .

"The enabling legislation for the NACA slipped through almost unnoticed as a rider attached to the Naval Appropriation Bill, on 3 March 1915. It was a traditional example of American polit-ical compromise. . . .

"For fiscal 1915, the fledgling organization received a budget of $5,000, an annual appro-priation that remained constant for the next five years."

NASA was created for the purpose of immediately correcting the gap between U.S. and Soviet launching, satellite, and general aerospace tech-nologies. The Advanced Research Projects Agency (ARPA) in the Depart-ment of Defense was created to insure that the U.S. would not lag in defense-critical technologies. These two organizations worked together in order to close the immediate gap demonstrated by the Sputnik incident.

Table 2.4   Indicators of Change in University Research

	Gross National Product	National Basic Research	% GNP	Total Univ. R&D	% GNP	Basic Univ. Research	% Nat. Basic Research	Federal % Univ. Research
1953	364,900	441	.12	255	.07	110	25	43
1960	506,500	1,197	.24	646	.12	433	36	67
1964	637,700	2,289	.36	1,275	.20	1,003	44	79
1968	873,400	3,296	.38	2,149	.25	1,649	50	77
1986	4,291,000	14,163	.33	10,600	.24	7,100	50	67

SOURCE: Geiger, Roger (1993)

NASA quickly became a player in funding academic research, buying its way into the universities in order to acquire access to relevant expertise. The funding of space sciences served as a complement to the massive space technology efforts undertaken to overcome the perceived gap between Soviet and American space capabilities. The growth of NASAs commitment to basic research boosted the academic science enterprise between 1960 and 1968 (Geiger, 1993). The post-Sputnik transformation of the university was a function of this and other military growth (Graham and Diamond, 1997).

ARPA, often in collaboration with the services, began supporting technology projects that were deemed critical to the long-term needs of the U.S. defense technology. ARPA did this by funding project-focused research within and across universities, companies, and government laboratories. ARPA maintained and managed an astounding portfolio of projects. ARPA's mission led it to mobilize all necessary resources to successfully meet the military's technological needs.

Perhaps the most overwhelming contribution ARPA made to the U.S. NIS beyond the creation of particular defense capabilities, was the creation of university and industrial capabilities in C3 technologies, technologies encompassing command, control, and communications. Various machines for data processing, computer languages, transmission hardware, satellite technologies, encryption software, networking technologies, and much more were developed under the coordination of ARPA research resources, providing significant incentives for the development of semiconductor technology (Nelson, 1962). Computer science departments were created in response to these new needs (Geiger, 1993). For longer term projects and for the development of discrete weapons systems labs were fostered such as MIT's Lincoln Labs (Leslie, 1993). This complex of technologies, held by or accessible to a population of government contract firms, served as the seedbed for innovation in commercial applications of the technologies. At the core of these innovations were labs that received significant support from public sources. These labs were supported by the defense establishment and were at universities, firms, and government sites. The demonstration of commercial use for these technologies created a market-based funding source for an explosion in lab-based innovative activity as new firms emerged to construct and seize new technologies and markets.

Lab Window 2.5: "Lincoln Labs: A Half Century of Federally Funded Cutting-Edge Electronics R&D"

Lincoln Labs

http://www.ll.mit.edu

HISTORY

Building on digital technology from the MIT Whirlwind Computer of the late 1940s, early research at Lincoln was focused on the design and prototype development of a network of ground-based radars and aircraft control centers for continental air defense. Significant technical advances supporting this work included the first real-time processing of radar data and the development of magnetic-core data storage to greatly increase system reliability and computer memory. Many of these technical developments later evolved into improved systems for the airborne detection, tracking and identification of aircraft and ground vehicles and formed the basis for research in space-based radar systems.

The Laboratory's Millstone Hill radar, completed in 1957, utilized the first all solid state, programmable digital computer for the real-time tracking of objects in space. In addition to its role in developing technology basic to the ballistic missile early warning system, the Millstone Hill facility was the first to track the Soviet Sputnik I satellite and served as a tracking station for Cape Canaveral launches.

Early laboratory studies of the properties of re-entry vehicles and the associated problems of radar detection, tracking and target discrimination were also instrumental in the development of ballistic missile defense strategies. In the early 1960s, Lincoln laboratory initiated a unique development program in satellite communications systems for national defense, and it has been successful in the design, construction, and launching of eight experimental active satellites. The Laboratory has demonstrated significant advances in autonomous spacecraft control, the use of solid state devices to ensure long-term reliability, and the development of mobile earth terminals for secure communications systems. The Laboratory has performed notable research in precision pointing and tracking of satellites using lasers, and has accomplished very significant experiments in compensating for the effects of atmospheric turbulence by using adaptive optics.

Since the early 1970s, Lincoln Laboratory has had an active program in civil air traffic control emphasizing radar surveillance, collision avoidance, hazardous weather detection and the use of automation aids in the control of aircraft.

To support its aggressive approach to advanced systems development, Lincoln has also maintained a leadership role in basic research in surface and solid state physics and materials relevant to solid state physics. The Laboratory performed the initial research for the development of the semi-conducting laser and designed an infrared laser radar to develop techniques for high-precision pointing and tracking of satellites.

Lincoln has also made significant contributions to the early development of modern computer graphics, the theory of digital signal processing, and the design and construction of the first high-speed digital signal processing computer.

Today, Lincoln Laboratory continues to be a primary source of technological innovation for advanced electronic systems.

FIELDS OF SPECIALIZATION

Air Traffic Control Systems

Communications Systems

Laser and Optical Systems

Radar Systems

Space Surveillance

Surface Surveillance

The blurring of the lines between R&D providers' sectors is apparent when one looks at the federally funded performers of R&D between 1967 and 1986 (table 2.5).

Energy Crisis and the Restructuring of Energy Research

During the period between 1966 and 1980, which has been termed a period of policy disarray, a number of issues that were not a part of the post–World War II consensus emerged (Geiger, 1993). Questions arose relating sluggish economic performance to the need for federal support of civilian technology. Other questions arose about the proper role of the military in performing military R&D in universities. The benevolence of science began to be questioned. Was science responsible for the many undesirable social spillovers from new technology? Compounding all this was the energy crisis and calls

Table 2.5 | **Federally Funded R&D, by Type of Performer, Selected Years, 1967–86**

Type of Performer	1967 Billions	%	1971 Billions	%	1976 Billions	%	1986 Billions	%
Federal agency intramural	3.4	23	4.2	28	5.6	28	13.5	24
Industry	8.4	58	7.7	52	10.2	51	30.0	52
Universities	1.4	10	1.7	11	2.5	12	8.1	14
Federally funded R&D centers	0.7	5	0.7	5	1.1	5	3.6	6
Other nonprofit institutions	0.6	4	0.7	5	0.8	4	2.1	4
Total	14.5	100	15.0	100	20.2	100	57.3	100

SOURCE: Smith (1990)

for the reorganization of federal R&D assets to deal with the nation's energy needs. The AEC investments in a strong set of multiprogram laboratories were scrutinized, and during this period the AEC and other national institutions dealing with energy questions were consolidated into a new Department of Energy (Brooks, 1996). Conflating national defense, environmental, and energy missions, the new energy organization passed on muddled, ill-defined missions to its labs.

Lab Window 2.6: "Industrial Spin-offs of Federal S&T Investment"

Lincoln Labs
http://www.ll.mit.edu/

Lincoln Laboratory has spun off more than 60 companies that employ 136,000 people and generate over $14 billion in sales; it has been granted 273 patents, 71 of which are licensed.

American Power Conversion Corp.	Janis Research Co., Inc.
Amtron Corp.	Jerome A. Lemieux
Applicon, Inc.	John Ackley Consultants
Arcon Corp.	Kopin Corp.
Ascension Technology, Inc.	Kulite Semiconductor Products, Inc.
Atlantic Aerospace Electronics Corp.	L. J. Ricardi, Inc.
Catalyst, Inc.	Laser Analytics, Inc.
Centocor, Inc.	Lasertron, Inc.
Clark, Rockoff and Associates	Louis Sutro Associates
Computer Corp. Of America	M. D. Field Co.
Corporate-Tech Planning, Inc.	Man Labs, Inc.
Delta Sciences	Meeks Associates, Inc.
Digital Computer Controls, Inc.	Metric Systems, Corp.
Digital Equipment Corp.	Micracor, Inc.
Electro-Optical Technology, Inc.	Micrilor, Inc.
Electronic Space Structures, Inc.	Micro-Bit Corp.
F.W.S. Engineering	MIT Francis Bitter National Magnet Laboratory
Gulf Coast Audio Design, Inc.	MITRE Corp.
Hermes Electronics, Inc.	Netexpress, Inc.
HH Controls Co., Inc.	Object Systems, Inc.
Hrand Saxenian Associates	Photon, Inc.
Information International, Inc.	Pugh-Roberts Associates, Inc.
Integrated Computing Engines, Inc.	Qei, Inc.
Interactive Data Corp.	RN Communications, Inc.

Schwartz Electro-Optics, Inc.	Transducer Products, Inc.
Signatron, Inc.	Technology Transfer Institute
sound/IMAGE	Tyco Laboratories, Inc.
Sparta, Inc.	U.S. Windpower, Inc.
Spiral Software Company	VIEWlogic Systems, Inc.
Synkinetics, Inc.	VV Imaging, Inc.
Tau-Tron, Inc.	Wolf Research & Development, Inc.
Telebyte Technology, Inc.	Xontech, Inc.
Telenet Communications, Inc.	Zeopower
Teratech Corp.	

Industries Spawned by Federal Funding of R&D

Even as the fruits of research are enjoyed, the ever-increasing dependence of American society on laboratory, science-based technologies is often ignored. Just as pre–World War II lab-based innovation depended on scientific and technical communities embedded in thick webs of supporting institutions, many recently emergent high-technology industrial sectors have their roots in government-supported institutions, institutions that could only exist with public support (Nelson, 1984).

The list of industries benefiting from public support for technology development is a long one. Public R&D support and procurement contributed to the development of semiconductors, supercomputers, computers, software, imaging, wireless communications, biotechnology, pharmaceuticals, aerospace, and virtually every form of energy technology from nuclear to passive solar. The technological skills and resources that industrial concerns currently utilize in their own industrial R&D activities have their roots in public support and are continually sustained by public support of graduate education.

Expanding the Nuclear Threat and Commercializing Nondefense Technologies

The ascendance of the Reagan administration was based, at least in part, on the belief that the U.S. was outmatched by foreign nuclear capabilities. Military spending ballooned, particularly on nuclear defenses, and military R&D grew as a result, though mostly on the development side (Brooks, 1996). The proportion of publicly funded R&D conducted for military purposes in industry, university, and government labs began to increase after a long decrease that had ensued after the elimination of the Sputnik technology gap.

Basic research fared reasonably well during the Reagan years when viewed within the context of the declining federal budgets. The basic science budget was held nearly flat, meaning a decline year by year due to inflation. But this was the fate of an aspect of research spending that was not ideologically opposed by the administration. Support for civilian technology programs was not even a possibility, as was evident in the zeroing out of efforts to identify government policies that could spur industrial innovation (Nelson, 1984). One exception was the creation of university-industry research centers by a number of agencies, centers that served as industry windows into academic research activities (Brooks, 1996). These centers were just one element of a set of policies receiving bipartisan support during the Reagan years, policies that dealt with the commercialization of technologies created by publicly funded research.

The first such policy was the Bayh-Dole Patent and Trademarks Amendment Act of 1980, which permitted contractors of federally funded research, including small businesses, nonprofits, and universities, to file for patents on and license, (on an exclusive or non-exclusive basis), the resulting inventions. Before the passage of Bayh-Dole, with some exceptions, the government's policy typically encouraged or required that such research results be placed in the public domain (Eisenberg, 1995). However, by the late 1970's, policy makers became concerned that this policy may, in fact, have impeded the commercial use of university inventions, noting that the government agencies, which retained title to the inventions, had little success in developing and marketing these inventions. Without exclusive rights to the inventions, the argument proceeded, private firms had little incentive to undertake the risk and cost associated with developing new products based on university research. By "solving" this problem, Bayh-Dole was seen as a means to enhance technology transfer and thus strengthen U.S. industry. The Stevenson-Wydler Amendments (called the Federal Technology Transfer Act of 1986) enabled federal laboratories to similarly transfer technologies through Cooperative Research and Development Agreements (CRADAs) with industry and other government agencies through integrated research programs, personnel exchange, lab space utilization, and research facility access (Branscomb, 1993). Technologies developed through any range of relationships with federal labs could then be the property of a firm seeking to commercialize them. The Federal Technology Transfer Act also provided for the enhancement of the Federal Laboratory Consortium for Technology Transfer that was initiated in 1971 to facilitate technology transfer across defense laboratories. The 1986 legislation required the development of stronger relationships between all federal labs and sought to help private sector agents to access and utilize federal R&D assets. Actions to free the private sector constraints on innovative activity included the R&D tax credit legislation of 1980 and joint R&D legislation of 1984.

Lab Window 2.7: "The Federal Blue Plate Special"

The Federal Laboratory Consortium for Technology Transfer
http://www.zyn.com/flc/theflc.htm

The Federal Laboratory Consortium for Technology Transfer (FLC) was organized in 1974 and formally chartered by the Federal Technology Transfer Act of 1986 to promote and to strengthen technology transfer nationwide. Today, more than 600 major federal laboratories and centers and their parent departments and agencies are FLC members.

The consortium creates an environment that adds value to and supports the technology transfer efforts of its members and potential partners. The FLC develops and tests transfer methods, addresses barriers to the process, provides training, highlights grassroots transfer efforts, and emphasizes national initiatives where technology transfer has a role. For thepublic and private sector, the FLC brings laboratories together with potential users of government-developed technologies. This is in part accomplished by the FLC Laboratory Locator Network and by regional and national meetings.

The purposes of the FLC are to:

- promote and facilitate the full range of technical cooperation between the federal laboratories and America's large and small businesses, academia, state and local governments, and federal agencies;
- provide direct services to member laboratories and agencies in support of their technology transfer efforts;
- enhance efforts that couple federal laboratories with American industry and small businesses to strengthen the nation's economic competitiveness;
- stimulate acceptance by the U.S. private and public sectors of the federal laboratory system and technology transfer as valuable assets;
- collaborate with local, state, regional, and national organizations that promote technical cooperation;
- promote further development and adoption of effective methods for federal laboratory domestic technology transfer;
- improve the effectiveness of individual and organizational efforts in technology transfer through training, recognition, awards, and evaluation of the FLC program;
- participate in and sponsor the development of novel models for federal laboratory cooperation with state, local, or private nonprofit technology transfer organizations; and
- serve as an interagency forum to develop and strengthen nationwide technology transfer in support of national policy.

This new commercialization emphasis placed new pressures on the federal laboratory system. This new mission, to be heeded by all major federal laboratories, potentially ran in direct opposition to other missions, particularly at labs engaged in basic research. With commercialization as a benchmark of success, many labs faced yet another pressure that aided in muddling their missions. New polices paid no heed to laboratories' or distinctive characters or capabilities.

Lab Science and Economic Competitiveness

The competitiveness debate spawned in the mid-1980s added a new chapter to the evolution of the U.S. lab system. The commercialization of technologies generated by federal funding was merely one sign of this, with the competitiveness discussion in the mid-1980s representing a more acute symptom. That debate was generally about our economic competitiveness vis-à-vis Japan and led those supporting governmental intervention to support some industrial policies (Johnson, 1984). At least some of the industrial policy activists connected waning competitiveness with inadequate technological innovation, especially in areas of "high technology." With protectionist viewpoints, such as advocating higher tariffs or similarly advantaging sunset industries, stumbling politically, the support of industries based on new technologies was seen as a viable alternative.

Elements of a technology policy that received bipartisan support during the Bush presidency included the Advanced Technology Program (ATP), the Manufacturing Extension Program (MEP), and the reorganization of the National Bureau of Standards into the National Institute for Standards and Technology (NIST) (Branscomb, 1993).

Lab Window 2.8: "Supporting American Technology Development for Nearly 100 Years"

National Institute of Standards and Technology
http://www.nist.gov

The National Institute of Standards and Technology (NIST), formerly the National Bureau of Standards (NBS), was established by Congress in 1901 to support industry, commerce, scientific institutions, and all branches of government. For nearly 100 years the NIST/NBS laboratories have worked with industry and government to advance measurement science and develop standards.

NBS was created at a time of enormous industrial development in the United States to help support the steel manufacturing, railroads, telephone, and electric power, all industries that were technically sophisticated for their time but lacked adequate standards. In creating NBS, Congress sought to address a long-standing need to provide standards of measurement for commerce and industry and support the "technology infrastructure" of the twentieth century.

In its first two decades, NBS won international recognition for its outstanding achievements in physical measurements, development of standards, and test methods—a tradition that has continued ever since. This early work laid the foundation for advances and improvements in many scientific and technical fields of the time, advances such as standards for lighting and electric power usage, temperature measurement of molten metals, and materials corrosion studies, testing, and metallurgy.

Both world wars found NBS deeply involved in mobilizing science to solve pressing weapons and war materials problems. After World War II, basic programs in nuclear and atomic physics, electronics, mathematics, computer research, and polymers as well as instrumentation, standards, and measurement research were instituted.

In the 1950s and 1960s, NBS research helped usher in the computer age and was employed in the space race after the stunning launch of Sputnik. The bureau's technical expertise led to assignments in the social concerns of the sixties: the environment and health and safety among others. By the seventies, energy conservation and fire research had also taken their place at NBS. The mid-to-late 1970s and 1980s found NBS returning with renewed vigor to its original mission focus in support of industry. In particular, increased emphasis was placed on addressing measurement problems in the emerging technologies. Many believe that the Stevenson-Wydler Act implemented throughout the federal laboratories the practices that had been developed at NBS over the years: cooperative research and technology transfer activities.

The Omnibus Trade and Competitiveness Act of 1988—in conjunction with 1987 legislation—augmented the institute's uniquely orchestrated customer-driven, laboratory-based research program aimed at enhancing the competitiveness of American industry by creating new program elements designed to help industry speed the commercialization of new technology. To reflect the agency's broader mission, the name was changed to the National Institute of Standards and Technology (NIST).

These efforts and the organizational changes brought about by the NIST Authorization Act for 1989 that created the Department of Commerce's Technology Administration to which NIST was transferred, contributes to a critical examination of the role of NIST in economic growth. These mission and organizational changes, initiated under the Bush administration, were reaffirmed and strengthened by the Clinton administration.

In addition to the reviews by Congress, the administration, and the Department of Commerce, the Visiting Committee on Advanced Technology (VCAT) of NIST reviews and makes recommendations regarding the general policy, organization, budget, and programs of NIST. The VCAT holds four business meetings each year with NIST management and summarizes its findings in an annual report submitted to the secretary of commerce and transmitted by the secretary to Congress.

ATP was created with the purpose of funding a class of research rarely undertaken by firms, research defined by high risk and high potential payoff—characteristics usually precluding market-based funding. This was seed money intended to help businesses largely focused on precompetitive, generic technologies with high market potential (Nelson, Peck, and Kalacheck, 1967; Branscomb, 1993). MEP was developed from the models and experience of the states of Georgia and Pennsylvania in order to support the innovative capacities of small and medium-sized manufacturers within a region. These means of support included individual project engineering, training courses, demonstrations, assistance in selecting software and equipment, services aiding innovation relationships, and the diffusion of the best practices. In many ways this MEP infrastructure had as its goal the diffusion of lab-based innovations to manufacturers and the support of manufacturers' ability to absorb and use such advances (Simons, 1993).

The Clinton administration's technology policy relied on ATP, MEP, and similar cooperative technology paradigms (Branscomb, 1993). Along with the congressional technology policy advocates, Clinton sought to expand the extent of these programs, dramatically increasing the funding available for ATP and the number of MEP manufacturing technology centers. These initiatives further complicated the landscape of laboratory policy by focusing on the aspects of science-based technological change that have largely been ignored by U.S. policy makers. Publicly supported policy has had its adherents since the Kennedy and Johnson administrations, but with little progress. Renewed calls for such action extended similar pressures to existing labs. Additionally, the Clinton-Gore science and technology policy strategy depended heavily on so-called dual-use technology programs as they sought to realign the defense R&D portion of the overall federal R&D portfolio. This policy shift in part attempted to create alternative policy rationales for national R&D assets that were previously supported because of the Cold War. Beyond the new rationale of "economic competitiveness" for defense R&D, this policy shift also helped to render the environments of many federal laboratories even more complex, presenting new demands in realms in which the labs had limited experience (Branscomb, 1993).

National Lab Reevaluation

From the beginning of the Clinton administration, the budget cutting environment forced a major stocktaking of all spending within the discretionary budget, including science and technology investments. One of the more vulnerable items within the federal science and technology budget was the system of national laboratories originally built by the AEC and now under the direction of the Department of Energy (DOE). Eight multiprogram laboratories requiring many billions of dollars in annual federal appropriations were again targeted for reevaluation, potential reprogramming, and waste cutting.

Earlier decisions by the secretary of energy to intensify the trend toward increased adoption of technology transfer activities and to make international economic competitiveness part of the mission of every DOE lab opened the labs to renewed analysis. The Galvin Commission, headed by Robert Galvin, the former chairman of Motorola, was created to undertake just this assessment. Although dealing only with the largest of the DOE labs, the findings of the report reflected the complicated political environment faced by all federal laboratories in the post–Cold War era. The report emphasized the potential resources these labs could offer the greater community and outlined the missions they should follow. These missions included maintaining the nation's basic science base, playing a role in supporting declining industrial research, serving as a technology transfer partner with industry and government agencies, dealing with local environmental problems, and supporting local education. Such a mission definition seems to provide little hope for a clear direction for DOE laboratories.

Laissez-Faire and Basic Science: Whither Applied?

The Republican capture of Congress in 1994 renewed momentum for laissez-faire notions of economic growth and limited government. With basic science viewed through a market-failure lens, investments in these fundamental science areas went largely untouched. At the same time, public investment in applied technology was subjected to attack, with the ATP serving as prime whipping boy. Most technology transfer programs were initially questioned. Even now the implications of the multiple mission of government labs performing basic and applied research as well as technology development have not played themselves out.

The Collapse of Corporate Central Research

The rise of commercialization and competitiveness missions in public laboratories is an adaptation to, but not a solution for the collapse of corporate central research. With the intensified globalization and deregulation of markets that sustained behemoth high-technology innovators like AT&T, GE, and IBM, the U.S. NIS has lost the presence of corporate central research laboratories that engage in long-term "directed basic." With these player-status private laboratories ceding their position as long-run investors in industry-relevant basic research, the innovation system is left with no institutions to fulfill this role (Rosenbloom and Spencer, 1996).

The new commercialization and competitiveness strategies suggested to public laboratories by policy makers are not geared toward supporting industry-specific, mission-oriented basic research. If the collapse of this particular mode of research in the U.S. NIS is considered an irreversible industrial trend, then the question of whether there is a need for public support of this mode of activity must be addressed. Merely muddling the missions of public laboratories with commercialization missions while failing to understand the diversity of science and technology investigations underpinning industrial success will contribute little.

Having no aspirations to add to the many fine histories of U.S. science and technology institutions and policy (e.g., Dupree, 1986; Smith, 1990; Hounshell, 1984), our more limited purpose in this chapter was to provide a background for an analysis of R&D laboratories in the U.S. Even from our brief account the basic character of the U.S. R&D enterprise emerges. One sees the diversity of U.S. R&D labs, their critical role in fundamentally supporting key sectors of our economy and society, and the interdependence of the innovation system's assets.

Our brief analysis of the R&D laboratories in the NIS draws from and at the same time reinforces the principles of the institutional design paradigm for science and technology policy. In line with the *systemic principle*, the overview examined the changing configurations of innovative activities in the U.S. NIS. History provides some of the central features of the change dynamics at work in the NIS. Changes in enabling technologies and in the NIS science base occur at every turn, and these as often reflect changes external to science. Sweeping changes in public educational institutions and concomitant changes in labor markets strongly affect the NIS. Epoch-making events such as depressions and wars affect the NIS, just as they affect virtually all of society's institutions. The effect of World War II on the structure of the NIS is nothing short of revolutionary; in some respects it led to the dawning of the consciousness of the NIS as an asset that can be systematically exploited in the service of national goals.

Before the twentieth century, politics and science rumbled down separate tracks. Gradually at first and then not so gradually after World War II, those tracks merged. For some time, the sensitivities of the U.S. NIS to large-scale political change have rendered it highly vulnerable. Among other results, political change often produces a "hiatus effect" (Link, Bozeman, and Crow, 1988). When science and technology policies take on the kind of on-again, off-again character produced by electoral swings, the need for systemic thinking becomes even more apparent.

Likewise, the *comparative advantage principle* stands out in any historical account of the NIS. Since the beginnings of the modern NIS, even the largest laboratories specialize. No laboratory does everything well. There seems to be no question about whether U.S. R&D laboratories play different roles, but much doubt concerning the integration of roles or the effects of one set of role-players on another. With the significant exception of the defense establishment, examples of successfully integrated roles are few. This should not be a surprise. Few coordinating mechanisms exist in an NIS dominated by a market failure paradigm for science and technology policy.

The *never-neutral principle* becomes especially important in the absence of coordinating mechanisms. During the early history of the NIS, government policies generally left a small footprint on the industry-dominated system. But today the footprint is much expanded. Government policies pervade all aspects of the NIS and shape its character, often through tax or regulatory policies as much as through science and technology policies. For example, any additional mission requirement placed on laboratories will in some way affect their performance. Merely commanding will not ensure performance. The complexity of the history and culture of various labs must be understood before new missions can be cast upon them.

Our minihistory's focus on the largest and most active of the laboratories is in accordance with the *player principle*. Realizing that many of the 16,000 laboratories in the American NIS are smaller science and engineering shops, one comes to understand that the design of government policies concerning labs should largely deal with high-capacity labs. In many respects, the player principle makes a systemic perspective possible. If the design task requires close attention to 16,000 laboratories, the task overwhelms. But keeping tabs on the players seems a much more manageable task.

Finally, the *opportunity cost principle* tells us to take into account alternative uses of limited resources. The benefits of investment must be judged against other available options. Any history of the U.S. NIS shows a transformation from small, decentralized efforts to dominance by large-scale, high-capacity laboratories. As the costs of big science rise, we find it impossible, even in the wealthiest of nations, to simultaneously support a human genome project, a space station, a supercollider, environmental remediation, and the next generation of weapons technology. Choices among science and technology options with high price tags are never easy. But increasingly science and technology options compete not only among themselves but with other costly social goals. The same trade-offs exist on a smaller scale within particular laboratories. Thus, diverting resources to technology transfer or mandating new economic development roles for labs often comes at the expense of existing activities.

Well into the 1990s the history of the U.S. NIS is one in which science and technology policies, including those directly affecting the R&D laboratory system, have developed in piecemeal fashion with relatively little system knowledge underpinning policies. The robustness of the 16,000 has permitted poorly integrated policies and limited coordinating institutions. But the NIS shows signs of strain. Resources are not so plentiful as to permit a continuation of unstable, poorly planned polices, many of which work at cross-purposes.

CHAPTER THREE

A Snapshot of U.S. R&D Laboratories in the National Innovation System

Structure, Output, and Design

In this chapter we begin our efforts to demystify the mysterious 16,000. Let us begin with an admonition. Science and technology policy makers would do well to act on the basis of what they think they know, rather than what they think but don't know. Of course, opinions and ideological predispositions are easier to come by than information. Execution of institutional design-based R&D policy making requires considerable information, including broad-based knowledge of R&D laboratory systems.

In chapter 1, we advanced several ideas as to why there is so little systemic knowledge about U.S. R&D labs, but the surface answer suffices: there have been few efforts to actually collect much systemic knowledge. In this chapter we provide an empirical snapshot of the R&D laboratory complex of the NIS. We give few recommendations. Instead, we seek to clarify issues about the composition of the R&D laboratory complex in the U.S., its structural design, and research orientation. The information provided in this chapter comes from several phases of the NCRDP. We use little information from other sources, not because we ignore information produced by others, but after an exhaustive search we finally concluded that no one else in the past twenty years has collected wide-ranging information on a large, representative sample of U.S. R&D laboratories (with Andrews [1979] possibly being the one exception).

U.S. R&D Laboratory System: Bringing the 16,000 Into the Light

During the 14-year course of the NCRDP, we and other project members have gathered data about U.S. R&D laboratories. It is neither practical nor desirable to provide in this chapter even a fraction of that information (for a more detailed account of aggregate statistical patterns for U.S. R&D labs see

Bozeman and Crow, 1988, 1990, 1995). We think it is useful, however, to present selected summary information about aggregate findings for the U.S. R&D laboratory system, chiefly because the NCRDP data is one of the few sources of information relating to some fundamental questions. Thus, this chapter presents highlights from several different phases of the NCRDP. Most of the results reported come from two phases of the NCRDP, data from the Phase II 1988 project (Bozeman and Crow, 1988) and data from a comparable 1991 effort (Crow and Bozeman, 1991) (see appendix 1). The first study includes data from 685 U.S. R&D laboratories of all types, including 120 of the largest 200 laboratories in the U.S. Whereas our 1988 study sought to represent the entire U.S. R&D laboratory complex, the 1991 study, while including many of the same respondents as the earlier study, focused on government laboratories. Hence, the second study is less representative of the U.S. R&D system. Whereas in the overall population of U.S. R&D laboratories, government labs represent only about 5%, fully 35% of the 1991 study respondents were government laboratories. Details about the methods and assumptions employed in NCRDP studies are available in the appendixes.

Profiling the U.S. R&D Laboratory System: The Sampling Population and Data Sources

By R&D laboratory system we mean the network of science-based, knowledge-producing organizations that has evolved in the U.S. over the last 100 years. These organizations exist within a highly interdependent set of relationships between bodies of scientific and technological knowledge and practice that depend on each other for knowledge feedstock, fundamental discoveries, and new technologies. These networks of R&D labs play key roles in a range of technological areas and in key areas of basic science as groupings of basic science labs evolve in a cooperative/competitive mode. Together these clusters and groupings of labs make up the U.S. R&D laboratory complex.

The data reported here are from various phases of the NCRDP, but chiefly from Phase II, a national survey. A first step involved defining this population of R&D laboratories. Four major research center directories were used to establish a population of U.S. R&D laboratories. Laboratories with less than 25 reported employees were excluded from the study population as were those chiefly conducting research in the social sciences. This yielded a study population of 16,597 R&D laboratories.

In drawing the sample for this study, both random probability and stratified sampling were used. A random probability sample of 1,300 labs was developed using a computer-generated random number list. In addition to random probability sampling, to assure representativeness it was deemed useful to gather information about the largest R&D laboratories in the U.S. The largest 200 laboratories (superlabs), as determined from an analysis of total laboratory personnel figures, were added to this list. After following the procedures detailed in appendix 1, the sampling population was trimmed to 1,341 and 966 responded to either a telephone or a mailed questionnaire instrument.

THE SUPERLABS

If one believes chapter 1's "player assumption"—that most R&D laboratories have little significance for the productive capacity of the NIS on the whole—then the interest in a purely representative depiction of R&D labs wanes. The modal R&D laboratory is a small, private company lab dominated by the company's short-term manufacturing and development agenda. In recognition of the need to understand the "players," the NCRDP data includes information about superlabs, the largest R&D laboratories in the U.S. Questionnaire-based information was sought from the 200 largest (in research personnel) labs and 123 provided information. Arguably, the information provided by the superlabs is particularly important inasmuch as they are all "players."

The superlabs are familiar. They have dominated American science and technology for more than 100 years. Some of these labs, such as Bell Labs, Westinghouse, Eastman Kodak, and IBM, have been players on a global scale during their entire existence. Others, such as Xerox Parc or the David Sarnoff Laboratory, are more recent constructs resulting from corporate changes or complete corporate divestitures. Others, such as Lawrence Berkeley Lab or Los Alamos National Laboratory, have been mainstays of the military/energy industrial complex since these communities of public and private interest evolved in the 1940s.

Lab Window 3.1: "From Corporate Central to Independent Lab"

Sarnoff Corporation
http://www.sarnoff.com/

PROFILE

Founded in 1942 as the RCA Laboratories, Sarnoff has been a wholly-owned, for-profit subsidiary of SRI International since 1987. Annual revenues exceed $120 million. President and CEO James E. Carnes leads a growing staff of 850. Sarnoff specializes in developing world-leading technologies and moving them rapidly into the marketplace by partnering with clients and investors. It works with clients in industry and government in areas that include:

Digital Television

As a member of the digital HDTV Grand Alliance, Sarnoff played a key role in developing and testing the technical standard for high-definition television. Sarnoff is also developing critical technologies needed to merge television and computers.

Computing
Efforts include a wide array of computer vision, parallel computing, and image processing applications. In its operation of the National Information Display Laboratory, Sarnoff is improving the accuracy of mammograms in the fight against breast cancer.

Biomedical
Using our expertise in microelectronics, Sarnoff is developing devices that will screen thousands of chemicals for possible use in new drugs and others that will quickly detect airborne bacteria.

Solid-State Electronics
Sarnoff is a custom designer and manufacturer of integrated circuits (ICs) for critical military uses and designs and develops visible and infrared focal-plane arrays.

Lab Window 3.2: "A Billion Here, A Billion There"

Los Alamos National Laboratory
http://www.lanl.gov

HISTORY
Los Alamos National Laboratory was established in 1943 as Project Y of the Manhattan Engineering District to develop the world's first atomic bomb under the leadership of J. Robert Oppenheimer. Today, Los Alamos is a multidisciplinary, multiprogram laboratory whose central mission still revolves around national security. Managed since its beginning by the University of California, where Oppenheimer was a professor, Los Alamos continues a commitment to maintaining a tradition of free inquiry and debate, which is essential to any scientific undertaking. The laboratory fills an intermediate role—between academic research and industrial production—that helps expedite the development and commercialization of emerging technologies.

MISSION
The laboratory's original mission to design, develop, and test nuclear weapons has broadened and evolved as technologies, U.S. priorities, and the world community have changed. Today, we use the core technical competencies developed for defense and civilian programs to carry out both our national security responsibilities and our broadly based programs in energy, nuclear safeguards, biomedical science, environmental protection and cleanup, computational science, materials science, and other basic sciences. The capabilities resident in these programs are increasingly being used in partnership with industrial firms to bring laboratory-developed technology to the assistance of the overall competitiveness of the U.S. economy.

Los Alamos National Laboratory was born with a compelling mission. Created during World War II as part of the Manhattan Project, our central mission originally was to develop the first atomic bombs under the wartime's tremendous sense of urgency.

After World War II our central mission evolved into developing nuclear weapons for deterrence during the Cold War. Today, the Cold War cycle of nuclear weapons development has ended, and so our new central mission must reflect the incredible global events of the recent past. In conjunction with the Department of Energy we define our primary mission as providing the technical foundation to reduce the global nuclear danger to ensure a more secure future for the nation.

Along with having a compelling mission, we focus on continuing to do great science in the service of the nation. By continuing to do great science—and many significant discoveries not directly related to national security have been made by Los Alamos researchers over the years—we will continue to be able to attract talented men and women to shore up our scientific and technical capabilities.

LOCATION
Located on the Pajarito Plateau about 35 miles northwest of Santa Fe, the capital of New Mexico. The laboratory covers more than 43 square miles of mesas and canyons in northern New Mexico.

STAFFING
As the largest institution and the largest employer in the area, the laboratory has approximately 7,000 University of California employees plus approximately 3,500 contractor personnel.

BUDGET
Annual budget is approximately $1 billion.

OWNERSHIP
DOE

BIRTH YEAR
1943

In aggregate, the superlabs represent more than 65% of the national research enterprise of more than $175–185 billion a year and the vast majority of scientific and technical personnel not on a university faculty and devoted to research. These labs are among the major players but not the "fastest" players in the U.S. R&D lab complex.

The Size and Distribution of Resources of the U.S. R&D Laboratory System

Our data on the size of U.S. R&D laboratories is heavily skewed by superlabs, which make up less than .5% of all laboratories operating in the U.S.

but are (by design) a much higher percentage of the laboratories in our data base. The median number of total employees for all responding laboratories is only 78, but such superlabs as Los Alamos National Lab, 3M, and IBM's Watson Research Center have several thousand employees each. Similarly, data on the total budget for the laboratories has the same skewed distribution. While the median budget is $2 million (1990 dollars), this disguises wide variation. The average budget of the bottom decile of laboratories is less than a few hundred thousand dollars, but the superlabs' budget averages more than $186 million (1990 dollars). The superlabs all have at least 1,000 personnel with some as large as almost 8,000.

The size of laboratories varies by sector, as is indicated by figure 3.1, which presents a pie chart displaying median total personnel for each sector. There is a much wider variance within sectors than among them. A rough measure of the administrative ratios for each sector can be derived by examining the total professional (i.e., scientific) personnel as a percentage of the total personnel. By this crude measure, the administrative component is somewhat greater in both universities (53.0%) and industry (56.4%) than in government (44.2%). Considering the superlabs as a group, the administrative ratio is identical to that of universities: 53.0%.

It is instructive to consider the laboratories in terms of the percentages of their budgets accounted for by government funding. For all laboratories in this sample, the mean percentage of government funding of R&D is 46%. However, it should be noted that there are many laboratories with 100% government funding and many with no government R&D funding whatsoever. Figures 3.2–3.4 are bar charts depicting the distribution of government funding in university, industry, and superlaboratories, respectively.

POLICY IMPLICATIONS: WHY PERSONNEL AND RESOURCES SIZE MATTERS

A number of policy questions pertain to size. Most of these questions have been posed before but never answered, at least not empirically. Indeed, the

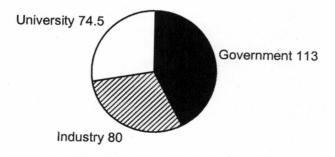

FIGURE 3.1   Median Total Personnel of Laboratories by Sector

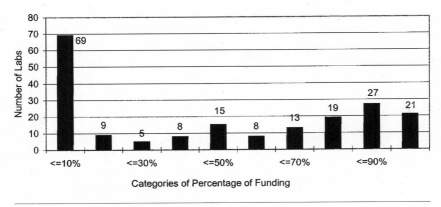

FIGURE 3.2 Distribution of Government Funding in University Labs

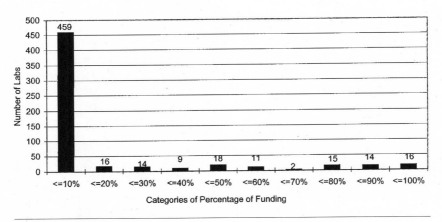

FIGURE 3.3 Distribution of Government Funding in Industrial Labs

issue of size effects on R&D policy is an interesting case study in surfeit and neglect. On the surfeit side, there are literally scores of studies on the relationship of the size of an organization to innovation (Kamien and Schwartz, 1982), studies showing conflicting and indeterminate findings. On the neglect side, there is almost no work on the impacts of R&D laboratories' size or resources. The lack of empirical knowledge of the size of R&D laboratories and effects of size is a critical gap in the information needed for R&D policy. The NCRDP data do not bridge that gap but at least begin to provide some descriptive information. Let us consider briefly some of the types of size-related questions that need answering.

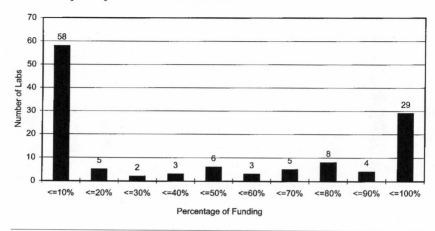

FIGURE 3.4   Distribution of Government Funding in Superlabs

A major gap in our knowledge for R&D policy: Are there size threshold effects that determine a lab's ability to make continuing, significant contributions to the NIS? No one knows the answer to this and to many other important size-related questions. On the one hand, the American myth of the small team of scientists or engineers convening in someone's garage and producing incredible, new, and rich technologies continues to have much appeal and at least some correspondence to reality. But these experiences are almost by definition heroic and uncommon. These stories of breakthroughs by lonely inventors probably have only limited relevance to today's high-technology, resources-based innovations. Even if one assumes, however, that the NIS is fueled chiefly by formalized R&D laboratories and corporate science and technology, the question "how much capacity is enough?" requires much more exploration. Our finding that 78 is the median number of scientific and technical personnel in R&D laboratories begs for perspective. Is there some size range at which labs become more effective? Do particular lab missions or scientific and technical agendas require more scientific and technical personnel than others? Are the effects of size mitigated by the lab's organizational structure? As lab directors and department heads seek to enhance innovation and at the same time to keep a lid on costs, questions relating size to scientific and technical productivity require more attention. But if such size questions are important at the management level, consider how much more important they are at the aggregate policy level. As various pundits provide confident prescriptions about downsizing or redeploying labs, they are doing so largely in the absence of any knowledge of scale effects.

We shall see in the next chapter that both scope and source of financial resources have an impact on laboratories' behaviors and output. But, again, many basic questions remain. Some arise almost automatically when one examines even rudimentary data. For example, figure 3.4 shows that, with respect to the superlabs, government financing of R&D is largely at the extremes: either government provides no or very little funding, or government provides almost all the laboratory's funding. What might occur if "mixed-type" superlabs were developed, ones expressly designed to be responsive to the market and also to have a significant public domain research role? While a handful of the largest labs already seem to play a mixed role, what might result should funding be used to reinforce such patterns? In fact, in Phase I of the NCRDP, we ran effectiveness assessments of various R&D lab types. The results of these assessments indicated that those R&D labs with balanced market and policy inputs and balanced science and technology focus were the most effective labs in operation. Resources questions such as these will emerge throughout the book.

RESEARCH ORIENTATIONS AT U.S. R&D LABORATORIES:
BASIC, APPLIED, AND DEVELOPMENT

Not only do U.S. R&D laboratories have distinctive research niches regarding fields of research and technology but also with respect to the orientation and overall purposes of the research. The classic basic/applied/development research distinction, while posing certain problems, is still a useful way of thinking about research orientation (Link, 1987, 1996). The focus in this brief overview is on the traditional tripartite distinction. However, the basic/applied/development categorization does not encompass the wide range of missions and objectives of R&D laboratories. Labs are involved in a wide range of missions including technology transfer, technical assistance, demonstration, and prototype development. The missions and research orientations of laboratories are considered in greater detail in a subsequent chapter.

Figure 3.5 is a chart indicating the percentage of laboratories (all laboratories, by sector) that view basic research as a major mission. These data confirm the expectation that universities dominate basic research. Perhaps the most interesting finding is that among the superlabs the focus on basic research is actually somewhat less than the average for all laboratories (this finding is largely explained by the fact that there are so few university labs among the superlabs). Figures 3.6 and 3.7 provide similar information for laboratories' orientations toward applied research and technology development, respectively.

Looking back at figure 3.5 and relating basic research performance to sector, we see that fully 31% of the respondents to this question view basic research as a major mission. As one might expect, there are substantial

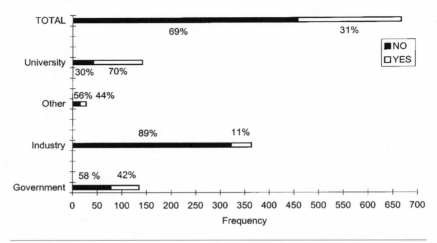

FIGURE 3.5   Basic Research

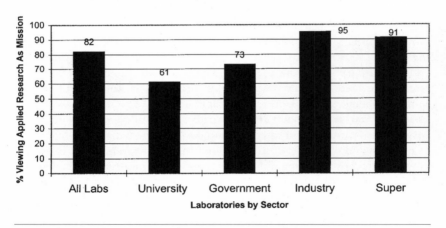

FIGURE 3.6   Laboratories that View Applied Research as a Major Mission

differences by sector, with 70% of all responding university laboratories involved in substantial basic research as opposed to 42% of government laboratories and 11% of industry laboratories. The trends are, then, in the direction one might expect. However, the sample includes some 41 industry laboratories involved in basic research as a "major mission" and, perhaps more surprising, 42 (30%) university laboratories that do not view basic research as a major part of their mission.

Understandably, laboratories oriented toward basic research are distinct

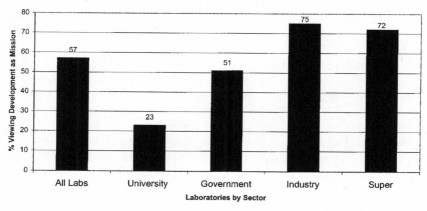

FIGURE 3.7   Laboratories that View Development as a Major Mission

from others. An examination of strongest correlates for the group of all laboratories (regardless of sector) oriented toward basic research reveals that these laboratories

1. have effectiveness criteria pertaining to scientific accomplishment,
2. have research articles and books and external conference papers as their principal output,
3. have a high level of government resources, government budgets, and government contracts,
4. are not oriented to market outputs,
5. are subject to government policies in the selection of research projects,
6. tend to be organized on a principal investigator basis (i.e., decentralized) and,
7. have as their most common evaluation mechanism evaluation by peers.

Thus, basic research laboratories are very much dependent on government support.

The pattern of applied research laboratories (figure 3.6) is easy to explain. Understandably, more than 95% of industry laboratories are engaged in applied research (and those that are not tend to be engaged in demonstration and organizational assistance, not basic research). Interestingly, there is little difference between government and university laboratories with respect to the prevalence of applied research.

An examination of correlates of applied research indicates that applied research organizations are in some respects the mirror image of those involved in basic research. The laboratories that are strongly oriented to market outputs, especially prototypes, are less likely to receive support from government, more likely to use commercial criteria as the chief effectiveness criterion, and, generally, have little interaction with government.

While there is relatively little difference between government and university laboratories with respect to the pervasiveness of applied research, government laboratories are much more likely (51% v. 23%) to be involved in the production of prototype devices. Most industry laboratories (75%) are engaged in the production of prototypes. The exceptions are chiefly comprised of industry laboratories with a high percentage of medical and biomedical personnel and laboratories that are largely technical assistance shops.

POLICY IMPLICATIONS: THE BASIC, APPLIED, DEVELOPMENT MIX

Interest in the composition of R&D is of long standing and, by and large, poorly served by studies of R&D laboratories. For years, the National Science Foundation in its *Science Indicators* (NSF, 1991, 1993, 1996) series has tried to conceptualize policy in terms of the composition of R&D and, especially, of spending on basic research. But most policy deliberations get no further than hand-wringing about whether or not there is a decline in basic research or whether industry R&D is too inward-looking. Once again, more sophisticated and strategically oriented questions are thwarted by a lack of empirical knowledge. While there is survey-based data on composition of

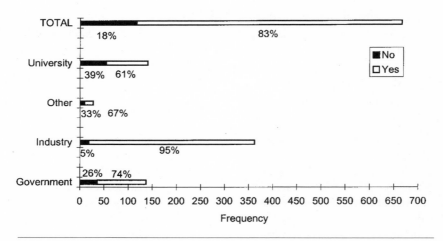

FIGURE 3.8   Applied Research

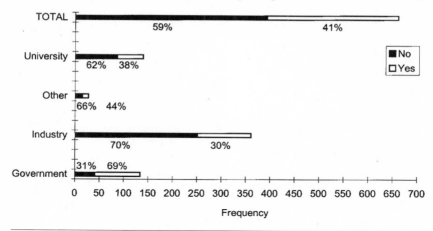

FIGURE 3.9   Technical Assistance to Government

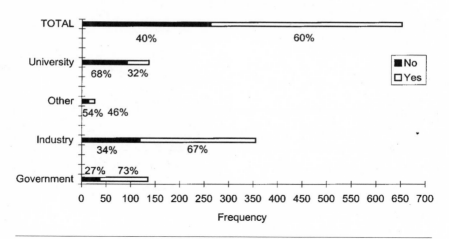

FIGURE 3.10   Technical Assistance to a Parent Organization

R&D spending (Battelle Memorial Institute, 1996), there is little or no effort to track shifts in labs' composition of R&D in connection with particular causes (e.g., alternative sources of financing) or effects (e.g., changes in innovation rates).

Economists have provided information about industrial firms' R&D mix and, in some instances, considered the effect of the mix by product or industry type. While this information is quite useful from the standpoint of the basic microeconomic model of the firm and has implications for managerial

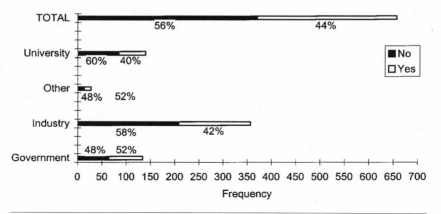

FIGURE 3.11   Technology Transfer to Commercial Organizations

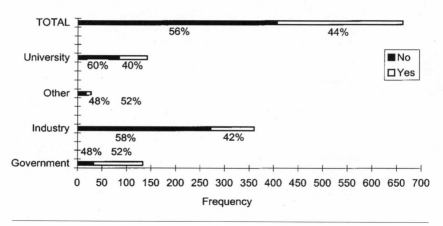

FIGURE 3.12   Technology Transfer to Government Organizations

strategy, there is little counterpart information for policy strategy. What is the mix of R&D in the U.S. R&D laboratory system and what difference does it make? The data presented above provides some rudimentary information about the "system mix" but no insight into the full implications of differences in the mix. While careful monitoring of the mix is an imposing task, keeping track of foci and shifts in the superlabs seems a less overwhelming task.

R&D LABORATORY OUTPUT

Some R&D laboratories devote most of their activity to the production of published scientific articles and books. Other laboratories focus on internal proprietary reports, and still others stress prototype development. Most laboratories have a considerable mix of activities, distinguished chiefly by their particular balance. Respondents were asked to report the percentage of laboratory activity devoted to each of several categories. As can be seen from figure 3.13, the figures for all laboratories mask a considerable variation among the sectors. Industry laboratories, for example, have quite limited involvement in the production of scientific articles, which is the mainstay of university laboratories. University laboratories are little involved in developing prototypes, whereas industry laboratories devote substantial time to prototypes.

Figure 3.14 provides similar output data for the superlabs. Interestingly, the distribution of activities for the superlabs is very close to that for the set of all laboratories. This is surprising when one considers that the sector representation of the superlabs is quite different from that for the entire sample. Almost no university labs are in the superlab sample, a disproportionate number of government labs are in the superlab sample, and industry labs are a smaller percentage than is found in the population or in our aggregate sample.

POLICY IMPLICATIONS: THE OUTPUT MIX

While R&D managers spend a good deal of time worrying about output mix in terms of both form and content, policy designers have given attention to

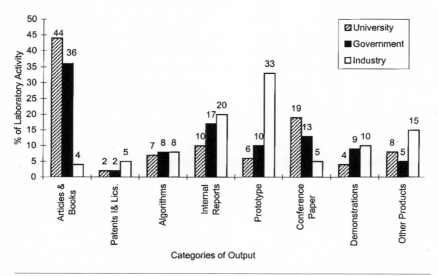

FIGURE 3.13   Percentage of Laboratory Activity Devoted to Categories of Output

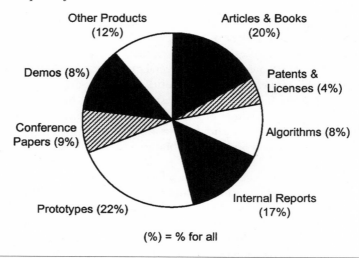

Other Products (12%)

Articles & Books (20%)

Demos (8%)

Patents & Licenses (4%)

Conference Papers (9%)

Algorithms (8%)

Prototypes (22%)

Internal Reports (17%)

(%) = % for all

FIGURE 3.14   Percentage of Laboratory Activity Devoted to Categories of Output for Superlabs

only the most obvious issues, such as the number of patents produced. Indeed, the output mix seems a problem for R&D managers rather than for R&D policy makers, but policy implications reside just below the surface. For example, one leading indicator of university labs' increased involvement in cooperative R&D and commercial technology may in general be a diminution of the percentage of output devoted to articles and books. Similarly, the federal labs' rates of patents and licensing are of interest as a gauge of the extent to which technology commercialization and transfer policies have taken hold. Thus, the output mix is not only an appropriate focus of policy strategy it may also prove to be a leading indicator.

SCIENTIFIC AND TECHNICAL PERSONNEL IN U.S. R&D LABORATORIES

As a result of various human resources studies sponsored by the National Science Foundation (National Science Foundation, 1990, 1993), scientific and technical personnel trends are among the few areas on which solid statistical information is available. Nevertheless, these data are not generally compiled on an aggregate R&D lab basis, and thus it is worth examining NCRDP data on personnel.

It is useful to consider not only the percentages of scientific and technical personnel by field among all R&D laboratories, but also the mix within laboratories. The "average mix," not reflected in any one laboratory, is presented in a pie chart in figure 3.15. While this average is helpful in understanding the distribution across all laboratories, individual laboratories tend to emphasize certain specialties to the exclusion of others. Only the superlabs have widely distributed personnel strength, and even among these laboratories there is considerable specialization. Figure 3.16, which focuses on the superlabs, gives both the mean personnel figure and the distribution of personnel

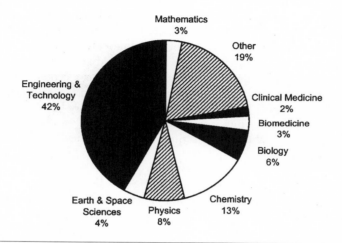

FIGURE 3.15   Percentage of Scientific and Technical Personnel by Area of Research

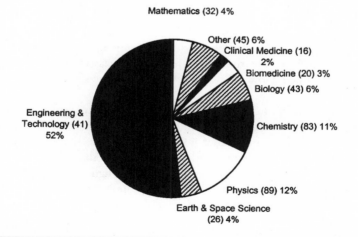

FIGURE 3.16   Mean Personnel and Percentage Distribution of Personnel by Area of Research

in percentages. This chart shows that the distribution for the superlabs is not unlike that for all laboratories. Perhaps the major distinction, and not a pronounced one, is the moderately higher percentage of physicists in the superlabs and the moderately lower percentage of chemists.

Engineering, chemistry, and physics dominate the technical enterprise among U.S. R&D laboratories. By contrast, scientific activity in the areas of astronomy is rare and such areas as mathematics, computer science, and environmental science are significant in number but are not widely distributed. Even allowing for the fact that sectors are unevenly represented in the sample, there is much difference in the research field concentrations in the respective sectors. For example, electrical engineering is a centerpiece activity of industrial R&D laboratories, but it is less common at government and university laboratories. There exists no industry laboratory built around the science of astronomy; while aeronautical engineering, underrepresented among the university laboratories, is quite a significant field among government laboratories.

Policy Implications: Scientific and Technical Personnel

As mentioned in chapter 1, considerable policy attention is devoted to the human producers and scientific field frameworks for analysis, and as a result there are data to inform strategic thinking. But in most instances policy issues are framed in market terms, with two questions at the forefront. First, is the production of scientists and technical personnel in a given field adequate for market demands? Is there an undersupply or oversupply of, say, aerospace engineers? Often these questions are framed in terms of projections, for example "given the growth rate of biotechnology firms is there likely to be a sufficient supply of geneticists, biologists, or medical researchers?" The second, related category of market-based questions often asked pertains to labor markets. What are the wages for various classes of scientific and technical personnel? How do wages affect initial career choice and career mobility?

Typically, little attention is given to human resources issues that are as much institutional as economic. It is important to know not only the dynamics of scientific and technical labor markets but also social and institutional factors. For example, aside from market pull, what are the "knowledge push" factors that draw scientific talent? How do breakthroughs and paradigm shifts in, say, astronomy affect the need for and the supply of astronomers? What drives people who choose math as their area of research? There are also "infrastructure pull" questions. For example, how does the development of new generations of critical scientific equipment affect the demand for technicians? Such questions force a perspective that harnesses market and institutional analysis.

ORGANIZATIONAL STRUCTURE AND DESIGN
WITHIN THE U.S. R&D LABORATORY SYSTEM

Among all policy questions relevant to strategic thinking about R&D policy, the ones most woefully neglected are those pertaining to organizational structure and design. There are several good reasons for this and at least one not so good reason. The good (or at least understandable) reason for neglect is that knowledge of structure and design absolutely requires conceptualization not at the usual individual level (e.g., individuals' career choices, individuals' wages for scientific work) or the sector level (e.g., patents in the steel industry, doctoral degrees produced by Research I universities) but at the less familiar laboratory or organizational level (i.e., the focus of NCRDP research).

The not-so-good reason for neglecting organizational structure and design issues is the erroneous view that structure and design issues are inherently the province of management, not policy. Strategic policy making requires knowledge of the organizational and institutional settings of R&D (and the focus of chapter 4 is entirely on this type of knowledge). In the absence of knowledge of institutional and organizational factors, one makes the potentially disastrous mistake of assuming that organizational characteristics are either minimally important or can be understood adequately in the broadest terms, such as simply considering whether a laboratory is owned by a university or by the federal government.

Institutional design issues are considered in the next chapter. Here we focus on structural variables. The structural variables fall into two categories. The first set of variables deals with the structure of organizations for research projects. Is research managed according to functional departments, units led by a principal investigator, or some other mechanism? The second category of structural variables concerns the bureaucratic structure. We examine a series of "bureaucratization" variables. Each of these is behaviorally anchored, providing information about the amount of time required for decision making within various R&D lab structures. While this is only one of many definitions of bureaucratization, using it is not an uncommon approach. Moreover, it is a particularly useful definition in light of the purposes of this study. As before, the analysis proceeds by examining the patterns for all laboratories and, then, by laboratories of each sector.

Typically, studies of organizational structure have sought one or both of two ends. Many have examined structure relative to performance (Dalton et. al, 1980; Bozeman, 1982). Others, especially those in the "contingency theory" tradition, have examined structural variables in relation to organizational environments (Ford and Slocum, 1977; Tolbert, 1985). Our concern is primarily with the latter. In part this is because no direct performance measures are included in this phase of NCRDP data. But, just as important, it is likely that performance is a function of the fit between structure and environment (Snow and Hrebiniak, 1980; Crow, 1985; Bozeman and Loveless, 1987). Thus, an enhanced understanding of the structure and environment of the U.S. R&D laboratory system should prove an important step in developing a more comprehensive understanding of performance.

UPON WHAT BASES ARE U.S. R&D LABORATORIES ORGANIZED?

Our concern here is not with the design of the entire organization but rather with the manner in which research projects are organized. We examined information as to whether the laboratory's research is organized into research groups led by a principal investigator, departments, divisions or branches, more or less ad hoc, based on the needs of the project, and "other." Many laboratories are organized on more than one basis.

Figure 3.17 presents the four types of research organization and the percentages of university, industry, and government laboratories employing them. The figure indicates that university laboratories are most often organized on a principal-investigator basis, government laboratories are most commonly organized on the basis of departments (but almost as often on the basis of principal investigator), and industry laboratories are often organized on the basis of department and principal investigator.

POLICY IMPLICATIONS: THE ORGANIZATIONAL STRUCTURE OF R&D LABS

Does the basis of organization relate to differences in laboratories' activities and characteristics? Apparently so. Our previous (Bozeman and Crow, 1988, 1995) examination of three-way contingency tables indicated that industry laboratories organized on a principal investigator basis are much more likely to be "public." These laboratories have higher percentages of government funds in relation to their total budgets, higher percentages of R&D funding from government sources, and are more likely to have direct R&D appropriations

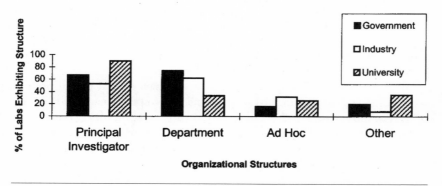

FIGURE 3.17   Organizational Structuring of Research in U.S. R&D Laboratories

from government and have higher percentages of their scientific equipment and facilities indirectly financed by government. Industry labs organized on this basis are also in closer contact with government via mail and telephone. These laboratories have a higher percentage of their R&D output in the form of articles, books, and external conference papers.

By contrast, industry laboratories organized on the basis of divisions and departments are less public in terms of the above factors, more likely to have a parent organization, rely more on their parent organization for their R&D funds, and most important, are much more likely to be among the largest industrial R&D organizations in terms of total personnel and total professional personnel (but not total budget). Industry laboratories organized on a departmental basis have a higher percentage of their R&D output directed toward patents. They have a higher percentage of chemists, medical, biological, and biomedical personnel among their scientific staff. Industrial laboratories organized on an ad hoc basis are smaller in every respect, less likely to have a parent organization, and most of their R&D product focuses on the production of internal technical reports.

Only the largest university laboratories are organized on a departmental basis (other than academic departments). They are likely to be among the largest 200 R&D laboratories, have high numbers of total personnel and total professional personnel, and large total budgets. Most are among the oldest university laboratories. These university laboratories have a high percentage of biologists, engineers, and physicists. Their R&D output orientation resembles that of other university laboratories.

University laboratories organized on an ad hoc basis differ chiefly with respect to their output, as these laboratories are much more likely to have as a high percentage of their output algorithms and software. Interestingly, those organized on an ad hoc basis tend to be somewhat more bureaucratic with respect to the amount of time required for approval to circulate research findings and the amount of time required for approval to publish findings. Characteristics of university labs organized by principal investigator are not considered because there is little variance as virtually all laboratories are to some extent organized on this basis.

Government laboratories organized on a principal investigator basis are quite similar to industry labs in that they are oriented toward output of scientific articles and conference papers. In other respects they are dissimilar. They tend to be among the smaller government labs in terms of total professional personnel and total personnel. There are no differences according to scientific fields of laboratory personnel. By contrast, government labs organized on a department or divisional basis are among the largest in terms of total personnel and total professional personnel, and indeed, many are among the largest 200 laboratories. Their predominant R&D output orientation is toward the demonstration of technological devices and the development of prototypes. They have disproportionate numbers of engineers and chemists. Government laboratories organized on an ad hoc basis are not distinct in any manner, except for a greater focus on algorithms and software.

The structuring of research is hardly random, but the direction of causality is not at all clear. It is not evident whether certain organizational bases develop as a result of organizations' environments or whether organizational structures permit movement into particular environments. In the absence of additional information, it seems likely that the former is more often the case, since it is apparent that laboratory administrators have some discretion in the choice of organizational basis.

WHAT ARE THE CHARACTERISTICS OF "BUREAUCRATIZED" R&D LABORATORIES?

In line with the assumption that organizational factors affect strategic R&D issues, much of the work of the NCRDP has focused on the prevalence and impact of "red tape" and bureaucratization of R&D laboratories (e.g., Crow and Bozeman, 1989a; Bozeman and Crow, 1991a; Cheng and Bozeman, 1993). One of the most useful measures of red tape used in the NCRDP as well as companion studies using data from a wide variety of organizations pertains to the amount of time required for each of a variety of policy and management actions (Rainey, Pandey, and Bozeman, 1995; Bozeman and Rainey, in press). The activities considered include:

- hiring full-time personnel
- hiring part-time personnel
- termination (because of poor performance or inadequate qualifications) of a full-time employee
- buying low-cost (less than $1,000) equipment
- buying expensive (more than $1,000) equipment
- submitting research results for publications
- circulating research results outside the laboratory
- getting internal funding for an individual investigator's research project
- getting approval for intermediate to large-scale team research projects

Table 3.1 shows the mean number of weeks for each activity by sector. The generalizations are easily uncovered. Government laboratories tend to be "more bureaucratic" on every factor (for elaboration see Crow and Bozeman, 1989a; Bozeman and Crow, 1995). For industry laboratories, tasks taking the longest time include getting internal approval and funding for medium- to

Table 3.1	Laboratory Red Tape: Average # of Weeks to Accomplish Tasks						
Ownership	Hire	Fire	Equip	Circ	Publish	Research	Total
Government	13.7	38.1	7.8	5.2	5.6	25.5	131.2
Industry	3.9	7.1	4.7	2.6	3.4	10.7	48.3
University	6.3	10.6	4.4	2.7	3.7	10.0	50.0
Other	3.9	6.2	4.9	1.5	2.3	6.0	35.9

large-risk team research projects. In government laboratories, research project approval takes considerable time but the dismissal of personnel consumes even more time. University and industry laboratories need about the same amount of time for research project approval. The pattern for the superlabs resembles that of industry laboratories except in dismissing personnel, and research project approval and funding takes more time.

Our earlier studies show that personnel and research project time relate closely to the degree of government funding (Bozeman and Crow, 1995). Size (in terms of total personnel) has little bearing on bureaucratic formalism. Our finding that size is not related to the degree of bureaucratic formalism is somewhat at odds with the general organization research literature (Kimberly, 1976; Pondy, 1969) but not at variance with previous findings for R&D laboratories (Marcson, 1960; Andrews, 1979).

Implications of Findings About the Structure of U.S. R&D Laboratories

Even the limited statistical profile we provide in this chapter offers some insights useful for framing policy deliberations about U.S. R&D labs. In this concluding section, we go beyond the data just presented to consider some of the ways in which the most basic attributes of the laboratory complex influence policy-making constraints and opportunities.

The infrastructure of the U.S. R&D laboratory enterprise is diverse. The system is composed of a set of organizations that are focused on particular classes of problems and functions. U.S. R &D laboratories focus chiefly on applied research and development. The system is dominated by small, private industrial laboratories with few connections to other institutions and organizations. The "player assumption" described in the first chapter—that most R&D laboratories have limited relevance for the NIS—seems borne out by the profile presented here.

The configuration of the U.S. R & D laboratories has some of the following general characteristics.

1. Knowledge production in the United States is not well organized or planned. There is no discernible pattern among the fields or areas of research that would appear to reflect a concern with optimized resource allocation. Given that only a small percentage of the resources are provided through competitive means, it is not possible to conclude that the best research is selected as a function of rational economic competition.

2. Knowledge is produced by a large number of individual providers. The universe of U.S. R&D laboratories exceeds 16,000. This number fluctuates as much as 3 to 5% in any given year. For the most part, these 16,000 R&D laboratories are small operations. Fewer than 30% have more than 100 employees, while many are comprised of a few engineers and technicians working on manufacturing support or a specialized technical product.

3. Organizational death seems to be commonplace in the R&D community. We base this conclusion on the fact that of the 1,341 laboratories initially examined, more than 100 had stopped independent operations either as a result of merger or the termination of operations. Since there was a three-year time period between our population source information and the time we called to verify operation, this suggests a 7% attrition rate for each three-year period. On the one hand, R&D laboratories are tenuous institutions, especially in the private sector. On the other hand, from a system perspective, the turnover rate implies potential for relatively short-term renewal.

4. The geographic concentration of the R&D laboratory system is largely in a few West Coast metropolitan areas, the Midwest, and the East Coast metroplex. This concentration pattern reflects manufacturing placement and federal R&D investment patterns.

What Is an Industrial Research Laboratory Today?

The modal industrial laboratory is easily described as a small, technically focused operation with little significance beyond one firm's short- and intermediate-term needs. But there are so many industrial laboratories that the category includes a wide array of R&D performers. Industrial labs range from technical services operations to multiple-capacity labs vital in providing innovation and public domain science and technology. As one would anticipate, industrial labs tend to be influenced heavily by the market, tend to produce proprietary research results, and for the most part have remarkably little direct interaction with R&D policy arms of the government (instead interaction is through the tax code and regulatory requirements). But the term "industrial lab" also includes market-based Federally Financed Research and Development Centers (FFRDC), entirely privately owned and many of them influenced strongly by government agencies and their agendas. Thus, despite the fact that as many as 90% of the 12,000 or so industrial labs are quite similar to one another in terms of scope and environmental interactions, the remaining thousand or so have great significance for the NIS.

The 1,000–2,000 industry labs that are "off type" differ from modal industry labs with respect to size and scale, mission, range of R&D output, and ties to other actors in the environment (particularly government actors). These are the dimensions that affect the capacity of labs, any labs, to contribute to the NIS. To be sure, small-scale industrial labs contribute to the U.S. economy by making the firms to which they are attached more productive. But their activities are best viewed as only marginally related to government policy because the *knowledge* externalities from these activities are usually quite modest. The "off-type" labs often have enormous scientific and technical capability and cannot conceivably appropriate all the knowledge they produce. They produce NIS-relevant spillovers as a by-product in many cases. These industrial laboratories are thus an appropriate target for strategic R&D policy. This does not, necessarily, imply a need for a high level of government intervention. It may simply mean that government should attend to the scientific health and welfare of industrial superlabs (and such attentiveness is commonplace, especially for defense-related critical technologies). Large-scale capacity is not the only attribute of industrial R&D labs that renders them relevant to policy. Many of the 1,000–2,000 NIS-relevant industrial labs are so laden with government contracts that they are in many respects little different in their missions from government laboratories. For these labs, different rules apply.

What Is a University Laboratory Today?

The NCRDP data focuses on autonomous and semi-autonomous university labs, ones not controlled by individual academic departments. This is not because the university chemistry lab is unimportant, but in many instances the correspondence between academic faculty units and labs is so close as to be inseparable. Moreover, recent years have witnessed a rise in the importance of university centers and multifocus laboratories. There are still hundreds of traditional university research groups led by a professor with a group of graduate students, post docs, and technicians, and these remain significant incubators of ideas.

The university research laboratories we concentrate on are organized research units, which are focused on either a class of problems or a general area of basic research. These organized research units are diverse and range from the NSF-funded Materials Research Labs to the larger DOE campus laboratories, such as Ames Laboratory (Iowa State University), Lawrence

Berkeley Laboratory (University of California), and the Radiation Lab (University of Notre Dame). We are also reminded of the NIH-supported medical schools and the huge labs that make up much of the biomedical research enterprise of the country. In addition, there are the agricultural experiment stations, with their wide range of funding sources, and the industrially funded, market-driven research institutes. This diversity in both size and function is further complicated by the complex character of government and market influence on these organizations.

Lab Window 3.3: "Church and State"

Notre Dame Radiation Laboratory
http://www.rad.nd.edu

Operated by the University of Notre Dame, Funded by DOE
FY 1994 Operating Budget $3,157,350
FY 1994 Staff 51

The Notre Dame Radiation Laboratory is the premier research laboratory in the United States for radiation chemistry, the study of chemical reactions induced by ionizing radiation.

The passage of radiation through a medium leaves a track of ions, unbound electrons, and radicals, which then react with molecules of the medium to produce the processes we study. Because of the close relationship between radiation chemistry and photochemistry (which initiates reactions with nonionizing radiation), a strong photochemistry program is both essential to and complementary to a strong radiation chemistry program. The effort is capped with a program of theoretical and computational chemistry that is closely integrated with the experimental investigations.

The central process studied in radiation chemistry is the one-electron transfer reaction. Reactions of this type are at the heart of almost every chemical device for the generation, storage, or transfer of energy. Although such processes can be studied by photochemistry or in thermal reactions, the characteristic production of free radicals in radiolysis gives radiation chemistry a substantial advantage over these other methods in pursuing electron transfer reactions.

The Notre Dame Radiation Laboratory is extremely well equipped for radiation chemistry and photochemistry, including several facilities that are unmatched elsewhere in the world.

Some university research laboratories are exclusively oriented toward basic research, others exclusively toward applied research. Almost all university labs are heavily financed by government; the exceptions are relatively uncommon university-based industrial service laboratories and the many mediocre university labs that do not compete well for government funding. A sizable portion of university laboratories is affected by a complex mix of government and market influences and do not conform to the university stereotype of pure basic research organizations. While it is easy enough to describe a modal industry lab, there is no template for university R&D laboratories.

University laboratories are extremely sensitive to changes in government policy and funding priorities. This means that university labs provide an important and accessible leverage point for policy change. But it also means that some diligence is required to ensure that sweeping changes in science and technology policy do not adversely affect university labs' contribution to the NIS. Sometimes fundamental changes in science and technology policy are not addressed to university labs but deeply affect them. Thus, for example, many university labs in the 1950s and 1960s became heavily defense-dependent. The downsizing of defense and defense R&D contracting in the early 1990s certainly was not initiated with the effects on university labs being uppermost in policy makers' minds, but the fact that the change was only "incidental" to the course of university research does not imply that it was insignificant. Many university labs found themselves frantically diversifying R&D portfolios, some with considerable success, some with little success. The "never neutral principle" articulated in chapter 1 applies in spades to university labs. Given the sensitivity of many university labs to changes in government priorities, deliberations about those priorities should include discussions of possible effects on the capacities and agendas of university labs.

What Is a Government Research Lab?

There is perhaps no greater mismatch between R&D perceptions and reality than in regard to government laboratories. The largest DOD and DOE multiprogram national labs dominate most policy makers' thinking about government R&D labs. Most assume that federal labs support the defense establishment and large-scale energy technology. But variety remains the hallmark of government labs.

Often, the ownership structure of government labs is quite complicated. First, many of the laboratories the government considers its own are actually operated under management contracts by various industries, universities, and private research institutes. These so-called GOCOs—government-owned, contractor-operated—constitute one of the major innovations in the institutional design of R&D laboratories. While many government laboratories directly serve their parent agency, many others have stronger ties to their clientele and users than to the government agencies. When adding to this mix the agricultural extension labs, the government-owned labs on university campuses, the DOD command labs, and the world-unto-themselves NIH labs, one readily sees the great variety in institutional design. If state governments are the "laboratories of democracy," government laboratories are the laboratories of design variation and innovation.

One of the most disappointing aspects of the recent round (1992–1997) of discussions about federal laboratories has been the extent to which discussions of lab downsizing or closure have so often failed to recognize the great institutional variation in government labs. The "systemic principle" demanding the evaluation of labs empirically and holistically can and should be applied in any policy making seeking a reconfigured federal laboratory system.

Despite diversity the broad outlines of the federal laboratory complex can be fathomed. While counts of federal labs vary, with about 500 on the low side and 800 on the high side (depending on how one deals with certain FFRDCs, small agriculture labs, and university-based federal research establishments), the "player principle" applies here just as elsewhere. Less than 10% of the federal labs account for more than 75% of scientific publications, patents, licenses, and R100 research awards. By developing a better understanding of about 100 federal labs, the role of federal labs in the NIS can be marked and tracked. While such an effort requires more than the usual government response of convening blue-ribbon panels, the level of effort needed to more fully understand these important contributors to the NIS is in all likelihood much less resource-intensive than what is required for one year's support of a small-scale technology development effort in one of the smallest, most obscure federal labs.

What Is the Possible Effect of Reliance on the Sector Model for Policy Making?

While this chapter overview is organized on the familiar basis of sector (university, industry, government), the NCRDP is motivated by the assumption that knowledge of the sector of a laboratory, in the absence of other institutional and environmental information, provides relatively little understanding. If federal policy makers wished to direct the federal laboratory network toward increased market relevance, this could result in dramatic changes and shifts in the output character of the R&D laboratory system. Reductions in the number of federal labs oriented toward the public domain could result in dramatic changes in the rate of production of public domain knowledge. These laboratories, largely because of their federal support, are focused on the production of knowledge that is important as an underpinning of the overall R&D system. These laboratories are in many cases heavily involved with others through cooperative arrangements. A change in the level of market influence on these organizations could result in a dramatic change in the overall character of the national capacity to develop new knowledge and technology.

The stereotypes, however, continue to prevail. For instance, many assume that universities perform most of the basic research in the country. What then are these university laboratories that perform much of the applied research in the country? For the most part, they are not the major autonomous university laboratories. Universities today live and die with their mission-oriented research programs. In addition, this assumption discounts the large-scale effort of a number of industry laboratories that are heavily financed by the government. That, of course, leaves the loosely organized university research group standing as major producer of basic research. This community of 75,000–100,000 researchers make up the cadre of knowledge creators. The steady drift toward centralized, free-standing university labs has strong effects on traditional research groups.

Sector affiliation is sometimes a good shorthand, especially in the absence of alternative ways of organizing knowledge. Sometimes crude, sector-based distinctions work well. Often they do not. In a nutshell, the chief advantage of the threefold, sector-based approach to thinking about R&D laboratories is its simplicity. The disadvantage is its likelihood of distorting the actual attributes and behavior of laboratories. In our view, sector-based knowledge is a beginning of R&D laboratory knowledge, but only a modest beginning.

If sector models have obvious defects, what alternatives best supplement the university-industry-government lab distinction? Obviously, the choice of analytical framework depends on the objective of analysis. If one's purpose is to understand laboratories so that bureaucratic controllers can make incremental changes, then agency affiliation seems a good place to start in analyzing federal laboratories. If one's purpose is to understand the health of the semiconductor industry, then it makes sense to focus on laboratories (of any sector) producing semiconductor technology and on the upstream industries using the technology. Indeed, these sorts of analytical perspectives have been used extensively and successfully.

Our focus is much broader than the laboratories of a government agency or a particular industry. We think these foci are crucial, but there is no conceptual gap hindering narrower-gauge analysis. Knowledge of components of the U.S. R&D laboratory system is available; it is knowledge of the system itself that is missing. In chapter 4, we turn our attention to an analytical framework that goes beyond consideration of sector affiliation and is much broader than any particular agency domain or industry. We seek to develop a model applicable to an analysis of the U.S. (or any nation's) R&D laboratory system. This model, which we term the "Environmental Context Taxonomy," focuses on the laboratories' niche within the overall system and positions laboratories in the system on the basis of their degrees of government and market influence. In our view, in addition to internal developments in science and technology, the evolution of R&D laboratories depends most strongly on market- and government-imposed constraints.

CHAPTER FOUR

The Environmental Context Taxonomy

A New Approach to Systemic Thinking About the U.S. National Innovation System

The senior senator from New Mexico buttonholes the chair of the House Science and Technology Policy Committee and fervently argues, "Personally, I think that there is a shortage of government-financing for hybrid science and technology labs because our appropriations are being frittered away on less productive public science and technology labs." The chair of the House Committee takes exception arguing "Wait a minute, as you well know the royalty revenue from the PS&Ts has been doubling every year and due to your party's recalcitrance the HSTs have begun to push the PS&Ts out of their market niche."

This fictional scenario is our way of saying that shorthand concepts will continue to dominate high-level debates for some time. It is difficult to imagine a scenario in which policy makers adopt the specialized taxonomy and resultant findings we present in this chapter. Policy debates are more likely to revolve around whether the DOE national labs should be shut down or whether university labs are providing an adequate supply of basic research.

Understandably, the "high game" debates require higher level and simpler concepts. But at the level just below the high game, science and technology policy specialists need information and conceptual tools well beyond those currently employed. If one is to understand the comparative advantage of R&D laboratories and their distinctive roles in the U.S. R&D laboratory system, one must have ways of thinking about institutions and environments.

While the resolution of conceptual problems concerning R&D laboratories is certainly not the only prerequisite for strategic R&D policy making (political will and a basic respect for strategy come to mind as others), these conceptual problems compose a set of puzzles attractive to science and technology policy analysts. Policy analysts can suggest that policy makers might serve all of us by one or another set of bold actions, but policy analysts' suggestions are consistently trumped by such powerful forces as political self-interest, public opinion, and empirically unfounded suggestions of policy makers. In many instances, the most valuable role for policy analysts is in developing new ways of thinking or new conceptual tools to deal with increasingly complex policy issues. Conceptual frames can have value even for those who do not agree with the policy analyst about policy directions, insofar as they offer a useful beginning point for analysis. Thus, concepts often last well beyond the latest, greatest policy crisis. In the case of R&D policy, there has been a dearth of new conceptual tools to help us to better understand these complex and socially important organizational structures. In this chapter we introduce frameworks developed in the NCRDP that present alternative ways of thinking about the U.S. (or any) R&D laboratory system.

The rationale for the development of new conceptual tools is that the R&D laboratory environment in the United States is one where the government, through direct funding, technology procurement, and public policy, has over the last 50 years been the single dominant force in the NIS. The result is that the simple model of sectoral-based laboratories (government, university, industry) and the resultant behavior expectations have been either weak in terms of their policy analysis utility or, in some instances, completely wrong.

Often government agencies provide the impetus for new institutional designs. Let us briefly consider a case in point. In the 1980s and early 1990s the Advanced Research Projects Agency (ARPA) created from whole cloth the parallel processing computing industry. Shaping this new advanced form of computing in its own image, ARPA set out to fund the major academic groups involved in systems development and design. ARPA initiated the procurement of prototype systems for military use and thereby stimulated the spin-off of several university groups into high tech start-up companies. These companies then set up R&D labs of their own that in turn were funded by ARPA. These firms began selling prototypes and second-generation machines to other ARPA research groups as well as to military and nonmilitary (but government-funded) research endeavors in the federal mission agencies. The result was the further evolution of university R&D labs to serve as the intellectual and financial incubators for a new technology needed by the military and, ultimately, the birth of a set of entirely new R&D-based companies in the field of parallel processing computation.

In this case ARPA realized the need for a new form of "supercomputing" and set out to create the R&D base and the technology development and transfer mechanisms to speed this technology to its desired ends. In so doing ARPA created, enhanced, or significantly influenced a number of R&D organizations. Through the deployment of its resources to directly manipulate the capabilities and performance of the NIS, ARPA drove the R&D process and the evolution of a new class of R&D organizations. Clearly these R&D

organizations were not operating in a simple sector-based model with defined roles where the universities provide basic research and industry provides technology development. Instead, a government agency husbanded resources from a variety of "players" in the NIS to, ultimately, create a set of new organizations. Conventional analyses of R&D policy making, focusing on conventional roles, could not adequately explain these major changes and the set of institutions and interorganizational ties that made them happen.

It is worth noting that since 1995 most of the new parallel processing computer companies started in the 1980s have gone out of business or have been acquired, for their technology only, by larger computer companies. These companies were early-stage innovators in a complex transformation period in computational design. Once it was demonstrated that the market for these specialty machines was, at least in this era, quite limited, the configuration of organizations and their research stabilized.

The complexity of the R&D laboratory system has grown over the last six decades as new R&D laboratory functions have evolved and, particularly, as networks of laboratories have become interrelated. The resulting complexity can best be illustrated by the changing roles of the CEOs of the research-intensive organizations.

Richard Atkinson and his "corporate" staff in their headquarters in Oakland, California, manage one of the most diverse and complex arrays of R&D laboratories in the world. With five major biomedical labs of at least 2,000 technical personnel each, two weapons labs with more than 5,000 employees each, a stand-alone basic science lab of more than 3,000 employees, the largest oceanic research institution in the world, the largest agricultural experiment station in the U.S., more than 500 organized research units of more than 25 personnel and billions of dollars in annual R&D funding, his job as the president of the University of California reflects the complexity of and rate of change in the R&D laboratory system. In 1940, virtually none of the major research entities that make up the university existed. Whole campuses such as UCLA or UC-Santa Barbara, now major research institutions in their own right, were either not yet even thought of or they were small extension sites for the main campus at Berkeley. The agricultural experiment station was less than 25% of its present size and the weapons labs had not yet even been conceived. The complexity of the relationships in this university holding company is best illustrated by the more than 2,000 technology licenses in force, tens of millions of dollars in annual industrial funding, billions in federal funding, hundreds of millions in state of California funding, and millions in foreign government funding fueling both the science and the complexity of the enterprise. One has to begin to seriously rethink any general conception of a university research laboratory when considering the role of Richard Atkinson.

Lab Window 4.1: "A Whole Lot of Laboratory"

Oak Ridge National Laboratory
http://www.ornl.gov

Oak Ridge National Laboratory, a U.S. Department of Energy facility managed by Lockheed Martin Energy Research Corporation, took root in an isolated East Tennessee valley during the Manhattan Project, the secret World War II race to develop the atomic bomb. When the war ended, ORNL turned its attention away from nuclear weaponry and toward the development of nuclear power and the production of radioisotopes for medicine and other peaceful purposes.

Over the years, as the nation's priorities have evolved, so have the laboratory's arenas of inquiry: New energy sources for the future. A more complete understanding of global warming. Better ways of handling toxic, hazardous, and radioactive wastes. Advanced materials crucial to industry. New tools to find and fight cancer.

Challenges such as these require the best minds and equipment science has to offer. To that end, ORNL employs close to 5,000 people full-time, including about 800 Ph.D.'s. Its scientific "toolbox" boasts particle accelerators that can build up or break down the fundamental units of matter, electron microscopes that can distinguish one atom from another, lasers that custom-make new materials one layer of molecules at a time, and one of the world's fastest supercomputers.

The laboratory is also home to 14 "user facilities" that are open to researchers from universities, industry, and other institutions throughout the world for studies of high-temperature ceramics and alloys, heavy-ion collisions, advanced building materials, large forest ecosystems, and other scientific phenomena. ORNL also maintains close ties with colleges and universities across the country. Each year, we host over 4,000 guest researchers from universities, other laboratories, and industry and more than 25,000 students from kindergarten through graduate school.

Unique equipment. World-class research facilities. First-rate scientific minds to guide the pursuit of knowledge and technological skill. These are the tools we offer at Oak Ridge National Laboratory.

Program areas include, physical sciences, environmental management, energy efficiency and renewable energy, biological and environmental research, defense programs, fusion, nuclear energy, fossil, environmental, safety, and health.

MISSION
The mission of the Oak Ridge National Laboratory is to conduct basic and applied research and development (R&D) in order to advance the nation's energy resources, environmental quality, and scientific knowledge and to contribute to educational foundations and national economic competitiveness. The Oak Ridge National Laboratory is committed to advance the frontiers of science and technology while addressing important national and global energy and environmental needs. The laboratory strives to be so well recognized for its excellence that students in certain fields of science and engineering regard working at the laboratory as an essential

element in their education. As a national laboratory, ORNL has an obligation to serve national needs through science and technology. In meeting this obligation, ORNL is committed to scientific and technical excellence and to contributing to the storehouse of basic scientific knowledge, using this knowledge to solve important problems and to create new opportunities, and attracting the best talent from all elements of American society. Intellectual honesty and scientific integrity are the foundation for this commitment.

STAFFING
ORNL employs close to 5,000 people full-time, including about 800 Ph.D.'s, 1,500 R&D professionals

BUDGET
Total FY 1995 Operating Cost: $545 Million

OWNERSHIP
Oak Ridge National Laboratory, a U.S. Department of Energy facility managed by Lockheed Martin Energy Research Corporation

---

In Bethesda, Maryland, just twenty minutes from the Capitol, sits Vance Coffman, CEO of the now largest defense contractor in the U.S., the Lockheed Martin Corporation. With acquisitions and consolidations, this complex organization now manages everything from Oak Ridge National Laboratory to R&D groups developing new air traffic control systems. Lockheed Martin runs some labs that are 90% funded for the public goods production, others that are 90% funded for classified science, with still others funded by corporate profits addressing a wide array of problems—making it a company technologically capable of dealing with many types of challenges. With hundreds of linkages to university labs, joint projects with government defense labs, project management and development responsibilities for NASA, and thousands of other relationships, the complex management responsibility of Coffman and his managers is orders of magnitude beyond the science and technology management responsibilities of the old Lockheed Corporation with its famous skunk works of the 1940s.

Not more than two miles from Vance Coffman sits Nobel laureate Harold Varmus who as the head of the National Institutes of Health manages the largest basic science enterprise on earth. Varmus's dominion includes a "central lab" of more than two dozen major institutes and centers and thousands of personnel within 50 major biomedical research centers run by universities but funded through NIH. Varmus manages a science and technology environment that is at the same time transforming U.S. demographics (through life extension), finding cures for cancer, mapping the human genome, tackling AIDS, and building new medical technologies. Through these science and technology efforts thousands of small R&D-based enterprises developed and hundreds of new R&D laboratories have emerged. The social, economic, ethical, and moral complexity facing Varmus and his research laboratories is well beyond that faced by most heads of state and is unique in both science and human history.

Why does one need more complex models for R&D policy? Because the R&D policy and institutions have become more complex. The time-honored classification tools cannot begin to capture this complexity. The world has a damnable tendency to ignore policy analysts' models and proceed to change entirely independent of those models. When this happens the analysts have the choice of failing with old, comfortable models or trying out new, sometimes less comfortable ones.

In Search of New Models:
Origins of the Environmental Context Taxonomy

In the spirit of trying new models when the old ones break down, we developed the *Environmental Context Taxonomy* (Crow, 1985; Bozeman and Crow, 1988, 1990; Crow and Bozeman, 1991; Bozeman and Crow, 1995). Most of this chapter is devoted to presenting and testing this model. The Environmental Context Taxonomy introduces a little more complexity and, we think, a good deal more understanding than one achieves using a sector-based (university-industry-government lab) framework. The taxonomy considers R&D laboratories in terms of two core dimensions, the laboratories' degree of *publicness* (Bozeman, 1987) and the economic character of the labs' output. But before outlining the model in some detail, let us consider further some of the changes that seem to us to require it.

There are two tap roots of the Environmental Context Taxonomy, one theory-based, the other purely empirical. The empirical base is derived from the first phase of the NCRDP, which consisted of a set of studies of energy technology labs (Crow, 1985; Crow and Bozeman, 1987b). Studies in this phase found that a significant percentage of labs defies classification in the standard nomenclature of industry-government-university. Site visits and case studies showed that as many as 30% of energy technology labs presented considerable difficulties in ownership classification. In some instances labs seemed to have hybrid forms; in other instances the ownership was clear enough but was mitigated by strong external control.

For instance, the Ames Laboratory of the U.S. Department of Energy in Ames, Iowa, has a staff of more than 500. These 500 are all employees of Iowa State University, the most senior of which, including the director, are tenured faculty members. The facilities are owned by the DOE but integrated into the campuses and physically linked to other university research facilities. What is this laboratory?

In part, empirical reality dictated the need for a new way of looking at R&D labs. But theory provided clues about a "new way." As a simultaneous development, we were developing a theory of organization seeking to show that the line between public and private organizations is more permeable than ever before (Bozeman, 1987). The term "publicness" was introduced into the literature on organization theory, signaling that political authority and political resources are exerted on all types of organizations (thus, the title *All Organizations Are Public*) and that "more public" business organizations are quite different from "less public" ones. The theoretical effort was joined to the empirical effort during the early phases of the NCRDP as we sought to test theory against the empirical reality of R&D laboratory attributes (e.g., Bozeman and Crow, 1990; Bozeman and Bretschneider, 1994).

In discussions of R&D laboratories, publicness refers to the degree of external control or influence exerted by governmental actors and includes such factors as the role of government in setting or shaping the laboratory's research agenda, the amount and percentage of laboratory resources that come from government (via contracts, grants, subsidies, or other such vehicles), and government control of the structuring and design of the lab. Importantly, the publicness concept does not equate with government ownership. It is even possible to think of industry-owned labs as more public than government-owned labs (though empirical instances are, of course, rare). Several studies (Crow and Bozeman, 1987b; Bozeman and Crow, 1995) have shown that the publicness of a laboratory often tells us much about its behavior and, related to that, that government involvement or control is almost never neutral in its effects.

The other primary classification factor examined with the taxonomy is the nature of the scientific and technological output of the lab, particularly the market or public characteristics of its outputs. To what extent is the lab oriented toward proprietary output, public domain output, or some mix of the two? The influence of the labs' knowledge products on its behavior has long been recognized but has not been well documented. This becomes particularly important in the policy environment of the late 1990s where there is a seemingly unceasing debate about the wisdom of government labs playing an increasing commercial role. But the other side of the issue is just as important: the increasing involvement of industry labs in producing public domain goods also leaves a mark on the character and culture of the industry lab (Bozeman and Crow, 1990).

As an illustration of the impact of publicness on private firms, consider again the case of the parallel processing computer companies that were stimulated by the actions of ARPA. Between the late 1980s and 1994–95 these new companies seemed destined for greatness. There were public offerings, major investments, and a general air of another American technological victory over perceived or potential external competitors. The need for such systems in the military was realized and many research groups around the country benefited from the new systems. After 1994–95, however, many of these companies collapsed. Market developments, particularly a gross miscalculation of the potential size of the market outside the military and the academic research community, led to failures. These failed companies led to the transfer of innovation leadership and market share in this field to the older, more established technology giants such as IBM. The complexities of market and public policy interaction proved to be difficult to manage for this group of new, relatively small R&D organizations.

This episode is quite suggestive for questions of strategic R&D policy. For instance, did this technology advance too quickly and without the kind of discipline that market forces place on commercial development? Was the government's role in helping to shape the R&D labs and the ultimate emergence of the companies a false market incentive or was it a standard government procurement effort for a new technology? Was the development of parallel processing technology ultimately set back as a result of the policy forces at work in this case, and if so, how could R&D policy makers and policy implementors better address such complexities?

The parallel processing case is one among many illustrating the sometimes unpredictable interplay between market and political forces across sectors of labs. But before introducing a set of concepts that cope with this reality better than traditional sector-based conceptual frames, it is useful to underscore that the traditional approach also has merit. As we have noted, our concern with hybrid organizations and the mix of economic and political factors leads us to seek new conceptualizations, but not to discard traditional ones that remain useful.

The Threefold Lab Stereotype: Advantages and Disadvantages

Stereotyping, in almost any instance, presents a convenient shorthand. The primary stereotype used in U.S. policy for R&D laboratories is the university-government-industry sectoral model. The advantages of the threefold stereotype are easily discerned. In the first place, it has an obvious validity. There are universities, and they do have labs. Likewise, some laboratories clearly are associated with government, others with industry. Just as important, policy makers are already accustomed to thinking in terms of these stereotypes. Other ways of thinking about laboratories (including our concepts) can be much more complex and can impede straightforward discussion of laboratories. The threefold stereotype clearly has its place. The preceding chapter provided some elementary descriptive information about laboratories, and it was useful to group the labs according to sector type. Later in the book we will focus on federal laboratories. While it is important to recognize the great variety among federal laboratories, it is just as important to understand that there are certain ways in which the behavior and status of federal laboratories is shaped by their ownership form.

The vignettes presented as prelude to the first chapter suggest the diversity of U.S. R&D labs, a diversity not fully captured by the three-sector model. Many laboratories are hybrids not easily labeled in terms of the usual three-sector distinction. As mentioned before, an NCRDP study found that of more than 200 energy-related R&D labs in the U.S. and Canada, fully one-third did not easily fit into a sector category (Crow and Bozeman, 1987a). Some of the most important laboratories in the U.S. are hybrids; the best known example being the DOE multiprogram government-owned, contractor-operated (GOCO) laboratories. But there are many less familiar examples of hybrid laboratories such as the human genome centers at universities and private institutes that receive core government funding and technology-specific industrial funding.

Lab Window 4.2: "R&D Joint Venturing"

SEMATECH

http://www.sematech.org

SEMATECH (SEmiconductor MAnufacturing TECHnology) is a nonprofit R&D consortium of U.S. semiconductor manufacturers. SEMATECH works with government and academia to sponsor and conduct research aimed at assuring leadership in semiconductor manufacturing technology for the U.S. semiconductor industry. It develops advanced semiconductor manufacturing processes, materials, and equipment and validates its research in a "proofing" facility that simulates manufacturing production lines. Results of this research are transferred to consortium members who use it for commercial and military applications. SEMATECH's emphasis is on manufacturing capability, not on device design. SEMATECH does not produce chips for sale.

SEMATECH members will create shared competitive advantage by working together to achieve and strengthen manufacturing technology leadership.

U.S. industry emerged from World War II as the most advanced and productive in the world. Since that time, our overseas competitors have, with their governments' encouragement and support, targeted specific industries for domination. The U.S. electronics industry, which invented semiconductors or computer chips, was one of those targeted. As a result, the U.S. lost leadership in the market and its ability to compete eroded.

Semiconductors are vital ingredients of practically every modern electronics system, from microwave ovens to communications satellites. They are the technological base of America's economic strength and defense strategy, which makes our ability to develop and competitively manufacture leading-edge designs critical to our nation's economy and security. Today, the U.S. industry has regained world leadership. SEMATECH played a supporting role in that rebounding of the U.S. industry. SEMATECH provides a cost-effective vehicle for developing advanced manufacturing technology that enables the U.S. to be competitive worldwide and to provide domestic, competitive sources of chips critical to the nation's defense.

In thinking about the complexity of the environment and the limits of sector-based classifications one might consider the Monsanto-funded laboratory at Washington University in St. Louis. The lab is the basic research and development center for the company in the area of biotechnology. Or consider the SEMATECH lab in Austin, Texas, funded by the government and a consortium of companies. The R&D universe is populated with many "off types," such as market-driven government labs (the agricultural product development labs of the USDA or the fishery research labs of the Department of Commerce), university labs that eschew basic research for product testing or technical assistance (there are more than 500 university labs dedicated to product testing and providing technical assistance), and industry labs that are world leaders in fundamental research (including Amgen and Biogen in the biotech sector, Lucent and IBM in the information and communications technology sector, and Illinois Superconductor in the emerging sector of superconductors). In some of the larger labs, the breadth of the R&D portfolio is such that no stereotype is likely to prove apt.

Forces for Change in R&D Laboratory Environments

In placid organizational environments, resources are localized, organizational actors are stable in structure and function, and there is relatively little interconnection of organizations, leaving autonomous actors to go their own way. Before World War II, most of the U.S. NIS operated in a placid environment characterized by low levels of governmental influence. Government activity was largely restricted to agricultural and public health research. Since that time, R&D laboratory environments have become increasingly complex and turbulent, characterized by diffuse resources, rapidly changing and unstable organizations, increased interorganizational dependency and reduced autonomy.

Looking back to about 1920, an era of placid environments for the R&D laboratories complex, the NIS was shaped by about 500 industrial laboratories, a handful of research universities, and a few dozen government laboratories, most of which were involved in agricultural research, weights and measures, or aviation-related research. The R&D laboratory complex was simple in its character and clear in its roles. With few exceptions, industrial laboratories pursued the production of knowledge for the short-term benefit of the parent firm. University laboratories focused on the training of advanced scientific and technical personnel. Government laboratories focused on particular technical issues related to the missions of parent agencies and, for the most part, these laboratories were either applied or directed basic research centers.

Between 1920 and 1929, as detailed in chapter 2, the U.S. R&D laboratory system reached a pinnacle, which some have termed the "golden age." During this time, U.S. R&D began to lead the world in its creative and productive outputs. In particular, the industrial R&D laboratory apparently had matured to the extent that organized research became as much a part of the modern corporation as accounting and finance. Labs at General Motors, DuPont, General Electric, Westinghouse, and AT&T became worldwide pacesetters in a relatively uncomplicated R&D environment characterized by specific technical goals, limited foreign competition, and labor-intensive science and technology. The development of new products and processes closely associated with the sponsoring firm's line of business would serve as the principal role of the industrial R&D laboratory for the next 40–50 years.

During this period, the roles of the various sectors operating in the U.S. economy seemed clear and in a sense became established in American culture. The U.S. government focused on a few technical service functions, such as the National Advisory Committee on Aeronautics and its airframe testing labs, or on assisting industries that lacked the market character to perform research such as the U.S. Bureau of Mines research stations or the agricultural experiment stations. In general the role of the government in the technical enterprise was supportive or ancillary to the industrial research enterprise. Industry was focused on market-oriented research, setting its own agenda and pursuing self-determined targets.

Universities, small by today's standards, were also supportive of the industrial research enterprise and began to expand their focus on fundamental science. Academic science was designed to underpin industrial development and economic competitiveness. Most university laboratories of that period operated in a placid environment and were organized around the individual research professor. The few university-based research bureaucracies were, for the most part, agricultural experiment stations and engineering experiment stations. Notable exceptions—and portents of the modern university R&D laboratory—were the Guggenheim Aviation Lab at Cal Tech (which served the aviation industry and later evolved into the Jet Propulsion Lab) and the university-based geological surveys, such as the famous Illinois Geological Survey in Urbana.

In spite of the economic dislocations of the 1930s, the environmental complexity of U.S. R&D laboratories had begun to stabilize and roles had begun to settle. World War II changed this. With the advent of World War II and the subsequent massive increase in the role of the government in both the funding of research and in its conduct, the U.S. R&D laboratory complex moved to a more turbulent environment. This new environment, which developed between 1942 and the late 1950s, significantly altered the roles of the university in the R&D laboratory system. The ad hoc structures and programs established during and just after World War II, such as the Manhattan Project, the Metallurgy Lab at the University of Chicago, the Radiation Lab at Notre Dame University, and the Applied Research Lab at Harvard (moved to Penn State in 1945) contributed to the successful conclusion of the war and to postwar technological advance.

Numerous new institutions were developed based on the assumption that the U.S. R&D laboratory complex would be able to use the resources not only to expand their roles but to ensure large-scale technological advance through increased federal funding. During and after the war, the modern, federally funded R&D establishment was born and changed the rules for science and technology in the U.S. (Bush, 1945).

Beginning in the late 1940s, the U.S. set out on a twenty-five-year building spree. This period included a tremendous expansion of research capability in terms of facilities and personnel and the development of a general structure for the management of segments of the still disjointed R&D laboratory system. Much of the R&D laboratory complex during this period was federalized and placed under the influence of research-coordinating agencies such as the National Science Foundation, the Atomic Energy Commission, the National Institutes of Health, and the Office of Naval Research. As a result, the environment external to the R&D laboratory complex became highly influenced by political authority. Up until this time, economic authority, operated through market push, was the dominant environmental force affecting U.S. R&D laboratories (Bozeman, 1987). As the federal government became the major force in the national research arena, technologies such as the integrated circuit, atomic power, polio vaccines, and advanced agricultural production were all viewed as products of the government's involvement in research.

Government involvement in the U.S. R&D laboratory system during this period resulted in numerous new types of organizations. For example, the Jet Propulsion Lab emerged as a major contract research facility of NASA, operated by the relatively tiny, elite California Institute of Technology. At about the same time, the GOCO laboratory type emerged as major federal research facility (e.g., Oak Ridge National Laboratory) and began to be operated by leading technology-based firms (e.g., Union Carbide and Martin Marietta).

With the greatly expanded government role, the sector-based roles within the R&D complex began to change. For example, MIT became the operator of one of the largest electronics laboratories in the world, Lincoln Lab. Funded largely by the Defense Department, the laboratory has operated as a major actor in the U.S. R&D complex for more than 35 years. The Lincoln Lab is a large-scale, mission-oriented research facility that is funded almost entirely by the federal government, conducted by a university-like entity, and operated, at least in part, to aid industry. This and other such institutions stand in stark contrast to the university's traditional department-based small research group led by a principal investigator.

During the 1960s, with still further governmental influence on the R&D laboratory system, the general environment became markedly turbulent. In addition to increased government influence, an explosion of new R&D laboratory units occurred in all sectors, many with unclear boundaries and missions. As a result of increased turbulence, the R&D laboratory complex began to show signs of serious stress by the 1970s. This stress seemed to manifest itself when R&D laboratories provided the technical fix for the energy crisis and other technology-relevant social problems. More important, however, the economy had begun to slow down. By the 1980s, the U.S. R&D laboratory system seemed less imposing, unable to provide technical solutions for any emergent problem, and new designs and new solutions began to be pursued (White House Science Council, 1983; Keyworth, 1983; Brooks and Schmitt, 1985; Roy and Shapley, 1985).

In the early 1980s, with economic and trade indicators reaching a critical stage and the R&D enterprise receiving increased scrutiny, the next large-scale environmental jolt to the R&D laboratory complex occurred. Several stocktaking efforts arrived at similar conclusions in diagnosing the problems of the U.S. R&D laboratory system (see especially the U.S. House of Representatives Committee on Science, Science Policy Task Force, 1987). One allegation was bureaucratic rigidity. Critics argued that R&D laboratories had become overly bureaucratized and no longer possessed the structural flexibility required for technological innovation. A second complaint alleged a lack of cooperation among institutions and sectors, inhibiting scientific and technical progress and commercialization of technology. Finally, evaluators alleged that the ambiguous missions and roles of R&D laboratories (especially government labs) led to a lack of direction. The "cooperative paradigm" (discussed in chapter 1) was invented (or, more accurately, imported) to address these perceived problems.

From the 1980s until now, several public policy options have emerged from the cooperative paradigm as well as several more in the tradition of the mission paradigm. Among the more prominent public policy responses of the past 15 years are the following:

1. Centralization of R&D planning and policy has been presented as a solution to perceived shortcomings of the NIS. One impetus was the perceived success of Japan's centralized approach to developing and commercializing new technology. There was a general belief that the U.S. lacked the planning and coordinating capacity to take advantage of the tremendous basic research and industrial development capacity available. Policy activity in this area has included the development of more government oversight for its wide-ranging research initiatives. This includes oversight of the mission agencies through new coordination mechanisms at the level of the executive office of the president. While not constituting centralized planning of the technical enterprise, this is a level of planned government oversight unseen in the past. As to how it has worked to date still remains to be seen. The enhancement in coordination and direction desired from the formation of the National Science and Technology Council in 1994 has yet to be realized. Nonetheless, it remains a policy objective.

2. Problem-oriented cooperative research centers have emerged as an approach to remedying the perceived gap between university research and industry needs. The currently popular dictum, "universities have departments, the real world has problems," implies that the university research community has problems with technical productivity and responsiveness. In addition, other cooperative ventures, such as the $200–250 million per year industry/government center in microelectronics known as SEMATECH, have over the last 10 years demonstrated the value of cooperation and coordination.

3. Refocusing government laboratories, especially the so-called national laboratories, toward more market-oriented research and enhanced technology transfer (Rahm, Bozeman and Crow, 1988) is a direct response to the series of criticisms presented in the Packard Report (White House Science Council, 1983). The role and function of the federal R&D enterprise is unclear to many, particularly in the context of industrial competitiveness. The Cooperative Research Act and the Federal Technology Transfer Act both expanded the federal laboratories' ability to contribute more directly to the commercialization of technology.

In the late 1990s, during yet another period of fundamental reexamination of the NIS, the initiatives developed under the cooperative paradigm have been called into question, but alternative policies have not yet supplanted these approaches. Presumably, market failure policies mean something more than reducing or eliminating policies born of cooperative or mission paradigm thinking. Historically, market failure policies have led to tax expenditure solutions or deregulation. But so far, the vortex of political change in the mid-1990s has not reformulated science and technology policy, it has simply dismantled it. Thus the impact government will have on U.S. R&D laboratories seems to be in a period of transition.

If Not the Threefold Stereotype, Then What?

The primary stereotype used in U.S. policy for R&D laboratories is the university-government-industry sectoral model. As discussed above, there are long-standing stereotypical roles ascribed to the labs of the three respective sectors: universities are seen as the bastion of fundamental research, industry as home of commercially motivated research and development work, and government labs are usually viewed as sites for supporting national research missions, especially in weapons, agriculture, and energy.

Many modern laboratories are hybrids that cannot easily be labeled in terms of the usual three-sector model. The best known among many examples is the DOE multiprogram GOCO lab, like the Ames Laboratory detailed above, that is government-owned and contractor-operated. In light of rapid environmental change and the limitations of the threefold stereotype of government, university, and industry labs, it is time to rethink conventional categories. But where to begin?

The Environmental Context Taxonomy

In developing the Environmental Context Taxonomy we sought to capture more dimensions of R&D laboratory environments than is possible in the traditional sector approach. Our thinking and our ultimate typology is premised on the assumption that laboratories, like other organizations, are influenced by a mix of economic and political authority (Bozeman, 1987). For R&D laboratories, environmental determinants include a variety of factors related to (1) the source and direction of resources imported into the laboratory and (2) the type and direction of R&D output emanating from the laboratory. Thus a typology is presented that has two coordinates: market influence and political/governmental influence. But prior to specifying the approach, it is useful to position it in the context of related works dealing with the conceptualization and measurement of organizational environments.

The R&D laboratory complex is ripe for taxonomic analysis. Since the late 1960s, many new organizational forms for R&D laboratories have developed (Teich and Lambright, 1976; Pavitt, 1984). These new organizational forms include: the cooperative research center (Norris, 1985; National Academy of Sciences, 1986; Mayfield and Schutzman, 1987), the joint R&D initiatives (Peck, 1986), the on-campus industry research center (Crow, 1985), and other new efforts at "federal laboratories" (Stark, 1984). All of these forms are new responses to the environment of the period. None fit neatly into government-industry-university pigeonholes.

In constructing a typology we build on earlier NCRDP work (Crow and Bozeman, 1987a, 1987b; Crow, 1985; Bozeman and Loveless, 1987), focusing chiefly on energy technology labs, to provide a typology applicable to the U.S. R&D complex. In this effort we address the following questions:

1. What are the major categories of R&D laboratories operating in the final days of the twentieth century?
2. What do the various categories of laboratories produce?
3. What are the structural and environmental characteristics of each laboratory category?
4. How do the various laboratory types respond to public policy initiatives?
5. Given certain policy or system goals, what types are best suited for production?

An overriding question is that of the effect of both market and public influences on laboratories of all types. By knowing how market and public constraints affect laboratories, critical issues concerning allocation of functions and responsibilities can be addressed.

The basic premise of environmental theories is that organizations adapt to their environment as a means of ensuring survival and that external influences are the dominant factors that affect these adaptations (Meyer et al., 1977; Aldrich, 1979; McKelvey, 1982). From this perspective, organizations can be seen not solely in terms of the behaviors and aspirations of individual decision makers but as components of complex interorganizational systems.

The changes that have occurred recently in the automotive industry offer a good illustration of this set of assumptions. By 1987, every major U.S. automobile manufacturer had in place a cooperative production and R&D arrangement with a major Japanese auto company. By 1997, every U.S. automobile manufacturer had linked new U.S.-based R&D centers with overseas R&D centers for "real-time" joint R&D.

By 1995, all U.S. automobile manufacturers had joined together with the government to form an electric vehicle consortium designed to address critical technology needs in battery, materials, systems, and engineering technology. This joint effort was developed in spite of the fact that GM, ahead of the others, was already moving forward with production of a 1997 model high-performance electric car, now for lease in the U.S.

In some cases, this reorganization of industrial R&D has resulted in the development and manufacture of new products overseas and in some cases it has resulted in new product development in the U.S. The reasons for this cooperative activity are diverse, but they can be meaningfully interpreted as adaptations for survival. As the environment for the sale of automobiles became more competitive, the two large forces, the U.S. and Japan, realized that there might be certain organizational and technological advantages to joint efforts. These advantages included: technology transfer, market dominance over European firms and other emerging Asian firms, and lower costs of production and new product development. The advantages vary among the actors, but the solution is the same: adaptation.

Another large-scale example of this process is the development of the Saturn Corporation by General Motors in the 1980s as the first integrated, computer-based manufacturing firm. In this case, General Motors, through its large capital base, attempted to develop an entirely new manufacturing paradigm within the firm. This revolutionary approach was undertaken as a mechanism for improved competition against the Japanese. The adaptations made here included computer-based decision making, paperless design and development, robotics manufacturing, and other labor-reducing efforts aimed at lowering the cost of production. External changes in technology were the driving force behind this adaptation. The development of a new organizational form to capitalize on the technological developments provided the means for adaptation. From an R&D perspective these adaptations also meant an integration of the R&D process with manufacturing to an extent previously not undertaken at General Motors. This exercise impelled by science and technology and management turned full circle in 1997 with the announcement by GM that a Saturn plant would soon be built in Japan.

Environmental forces influence the behavior of organizations so extensively that their performance and capability can be better understood through an assessment of their environmental niche. In the case of R&D organizations, as is true for many organizations, the dominant environmental influences are market and political/governmental forces. Both political and economic authority influence organizations with regard to purpose, ownership of property, structure, and decision calculus (Dahl and Lindblom, 1953). An organization heavily influenced by economic authority will tend to focus on economic self-interest and proprietary concerns. An organization influenced by political authority will tend to focus on the public domain, political agenda setting (with special attention to the political interests of its sponsor or parent), and the maintenance of a political economy. Political and economic rationality are not always at odds but often are countervailing influences on the behavior of organizations. Some of the "most public" organizations are, from an ownership perspective, private, such as Lockheed Martin and its research labs. And some of the most entrepreneurial and market-oriented or market-influenced R&D laboratories are government-owned, for instance, the citrus research centers of the Institute of Food and Agricultural Sciences at the University of Florida and the food product labs of the USDA throughout the U.S.

The Environmental Context Taxonomy, Phase I: Mapping Energy Labs

In Phase I of the NCRDP, an environmentally based taxonomy for the classification of R&D laboratories was developed (Crow, 1985). Based on the assumptions described above and presented in detail elsewhere (Crow and Bozeman, 1987a, 1987b; Crow, 1985), the taxonomy was configured around a nine-celled matrix. This classification matrix is a simple representation of the various levels of governmental and/or market influence that might affect the behavior of R&D laboratories.

From this matrix, with its simple presentation of levels of influence, several laboratory types were identified. This classification terminology was developed prior to the implementation of the Phase I classification exercise and reflects the character of the environmental setting. Each environmental setting is unique in the mix of influences and therefore is designed to capture similar organizations, regardless of ownership or other characteristics.

In Phase I of the NCRDP, this new classification typology for R&D laboratories was tested on a subset of the R&D system. We examined 829 energy-related R&D laboratories as an initial test of the typology (tables 4.1 and 4.2) (Crow, 1985). After determining a good deal of predictive value in the "cells" of our typology, we developed names for laboratory types, names that seemed to capture their character and function within the R&D system.

To implement the taxonomy in this first test exercise, simple indicators for both governmental and market influence were identified, such as percentage

Table 4.1		**Classification by Ownership of the Study Population**
Ownership Class	**Number of units**	**Category Description**
Industrial	258 (31.1%)	Laboratory facilities are owned by an industrial organization.
Government	74 (8.9%)	Laboratory facilities and programs are owned and operated by an agency of the government.
Other	497 (59.9%)	Laboratory facilities and programs are cooperatively owned and operated by more than one organizational owner.
Total	829 (99.9%)	

| Table 4.2 | **Ownership Distribution of Respondents** | | |

Ownership	Frequency	Percent	Cumulative Percent
Industrial	46	18.40%	18.40%
Govt. agency	41	16.40%	34.80%
Other	163	65.20%	100.00%
Total	250	100.00%	

of funding by source. Using these indicators, questionnaires were obtained from the 250 respondents (laboratory directors) resulting in the classifications shown in table 4.3. Further on-site analysis of the representative respondents indicated that the classification typology distinguished among laboratories in significant ways.

Although limited by the scope of the effort, just 250 laboratories, the Phase I classification exercise clearly indicated that the new R&D classification could provide insight into systems of R&D laboratories. An organizational effectiveness evaluation was carried out on 32 selected cases from various cells of the Phase I matrix.

Combined with our testing of the typology, our case studies showed that the two chief environmental dimensions we examined provided a great deal of information about laboratory structure and performance (Crow and Bozeman, 1987a, 1987b). But aside from the substantive findings from Phase I, its chief result was to give us sufficient encouragement to use this same typology to perform a more detailed analysis of a representative sample of the entire U.S. R&D laboratory system, detailed as Phase II of the NCRDP.

The Environmental Context Taxonomy, Phase II: Mapping the U.S. R&D Laboratory Complex

Phase II of the NCRDP used similar methods to test this new classification taxonomy for a fully representative sample of all R&D laboratories in the U.S. Given the discriminating power of the Phase I approach, we continued the basic design typology with two coordinates: market influence and political/government influence. However, additional measures for both influences were required.

GOVERNMENT INFLUENCE

To determine the level of public influence that might affect the behavior of an R&D laboratory, a new indicator of government resource dependence was developed. In Phase I, government influence was measured by one factor

Table 4.3	**Classification of R&D Organizations by Environmental Influence**			

Level of Market Influence	Level of Government Influence			
(nature of R&D products)	**High**	**Moderate**	**Low**	**Row Total**
Generic product (low)	Public Science	Hybrid Science	Private Niche Science	
Total	121	40	6	167
Row %	72.5	24	3.6	68.2
Column %	89.6	64.5	12.5	
Total %	49.4	16.3	2.4	
Balanced Product (moderate)	Public S&T	Hybrid S&T	Private S&T	
Total	4	11	10	25
Row %	16	44	20	10.2
Column %	3	17.7	20.8	
Total %	1.6	4.5	4.1	
Proprietary product (high)	Public Technology	Hybrid Technology	Private Technology	
Total	10	11	32	245
Row %	18.9	20.8	60.4	100
Column %	7.4	17.7	66.7	
Total %	4.1	4.5	13.1	
Column	135	62	32	
Total %	55.1	25.3	19.6	

Number of missing observations: 5

only: the percentage of government funding the company received. In Phase II, three government resource indicators were employed: percentage of R&D budget provided directly by the government, percentage of scientific equipment and facilities financed directly by the government, and the percentage of equipment and facilities financed indirectly by the government.

The assumption used for this classification is that resources are an important tool of government research policy (Bozeman, 1987). The extent to which any R&D laboratory utilizes these resources is an indicator of the level of influence that the government might have over the organization. The Phase I analysis indicates the plausibility of this assumption.

MARKET INFLUENCE

Market influence was determined by the extent to which respondents classified their laboratory effectiveness as being dependent upon either commercial or scientific success. We assumed laboratories measuring their effectiveness based on commercial success criteria have a higher level of market influence than those measuring effectiveness according to scientific performance.

The market variable places each laboratory on a discrete scale from total reliance on commercial success to total reliance on scientific success as an indicator of effectiveness. In this way each of the respondent laboratories was classified as having high, moderate, or low levels of market influence.

CLASSIFICATION IMPLEMENTATION

With these indicators, a Phase II typology was developed. Laboratory distribution according to this typology is presented in table 4.4.

With the classification design complete, the task of placing respondents within the matrix was carried out. Of the 966 respondents to the survey, 678 completed all of the questions necessary to be classified. The distribution of these 678 respondents according to the traditional sector characteristics is

Government =	97	14.3%
Industrial =	407	60.0%
University =	137	20.2%
Other =	37	5.5%

These levels roughly parallel the sector distribution in the overall sample and also the population of R&D laboratories operating in the U.S.

Laboratory Characteristics

Table 4.5 shows the distribution of US R&D laboratories according to taxonomic cell and sector. All cells except the private science laboratory have a

Table 4.4   **Laboratory Distribution of Taxonomic Cell**

Level of Market Influence	Level of Government Influence		
	Low	Moderate	High
Low	Private Niche Science ISD	Hybrid Science 54	Public Science 133
Moderate	Private Science and Technology 59	Hybrid Science and Technology 73	Public Science and Technology 77
High	Private Technology 175	Hybrid Technology 82	Public Technology 25

ISD = Insufficient Data

sufficient number of observations to permit analysis. That there are so few private niche science laboratories is probably due to the fact that virtually all large, complex organizations are substantially influenced by either the market or government. Those not strongly influenced by either are perhaps much like the "country-club" type of organizations identified by Blake and Mouton (1968). But whereas one would predict that loosely formed organizations with limited resources would not be effective, at least some of these organizations seem to prosper in finding narrow intellectual niches.

A prominent example of a private science laboratory is the Carnegie Institution in Washington with its physical science base and independent agenda-setting capability. It has demonstrated the ability to have a tremendous impact in the area of astronomy and the basic earth sciences.

Lab Window 4.3: "Private Niche Science"

Carnegie Institution of Washington
http://sak.ciw.edu/

HISTORY
THE CARNEGIE INSTITUTION OF WASHINGTON, a private, nonprofit organization engaged in basic research and advanced education in biology, astronomy, and the earth sciences, was founded by Andrew Carnegie in 1902 and incorporated by act of Congress in 1904. Mr. Carnegie, who provided an initial endowment of $10 million and later gave additional millions, conceived the institution's purpose as "to encourage, in the broadest and most liberal manner, investigation, research, and discovery, and the application of knowledge to the improvement of mankind."

From its earliest years, the Carnegie Institution has been a pioneering research organization, devoted to fields of inquiry that its trustees and staff consider among the most significant in the development of science and scholarship. Its funds are used primarily to support investigations at its own research departments. Recognizing that fundamental research is closely related to the development of outstanding young scholars, the institution conducts a strong program of advanced education at the predoctoral and postdoctoral levels. Carnegie also conducts distinctive programs for elementary school teachers and children in Washington, D.C. At First Light, a Saturday "hands-on" science school, elementary school students explore worlds within and around them. At summer sessions of the Carnegie Academy for Science Education, elementary school teachers learn interactive techniques of science teaching.

MISSION
The institution today operates five scientific departments having a total of about 60 staff scientists. A few senior scientists continue active research after official retirement. Each year about 120 fellows, associates, students, and visitors are in residence, so that the total academic membership, exclusive of supporting staff, is about 180. The institution's administrative offices, including the office of the president, are in downtown Washington, D.C. The five departments are:

The Department of Embryology in Baltimore, Maryland, whose investigators seek a better understanding of the molecular and cellular mechanisms underlying differentiation, growth, and morphogenesis and the manner in which these processes are coordinated in a number of developing systems;

The Department of Plant Biology in Stanford, California, whose scientists conduct research on physiological, biochemical, and genetic mechanisms underlying plant evolution and adaptation to different environments and apply advances in plant molecular biology to fundamental problems in growth and development;

The Geophysical Laboratory in Washington, D.C., where scientists conduct physicochemical studies of geological problems, with particular emphasis on the processes involved in the formation and evolution of the earth's crust, mantle, and core;

The Department of Terrestrial Magnetism in Washington, D.C., where the staff treats a wide range of topics in astrophysics, seismology, geochemistry, cosmochemistry, and planetary physics; and

The Observatories of the Carnegie Institution, based in Pasadena, California, where astronomers conduct programs of research on the structure and dimensions of the universe and the physical nature, chemical composition, and evolution of celestial objects.

RELATIONSHIP TO SECTOR

Recognizing that industrial organizations have public influences and government organizations have market influences does not imply that there is a random distribution of effects among the sectors. As expected, the classification matrix shows that organizations tend to cluster according to sector, though with some conspicuous exceptions.

Table 4.5 shows that industrial laboratories, while found in each cell of the matrix, are concentrated in cells 4, 5, 7, and 8. Of the industrial laboratories, over 41% can be classified as private technology laboratories and more than 30% as one of the other three types. Over 90% of government laboratories are pure public science or public science and technology labs; the remaining are spread out among cells 2, 8, 9. Of these, only the public technology laboratories in cell 9 are significantly made up of government-owned labs.

As might be expected, 70% of university laboratories are clustered in cells 2 and 3, but they are also found in nearly every other environmental setting. Likewise, laboratories whose sector class is "other" are most evident in cells 2 and 3, but they are also scattered among the other cells.

These classification results indicate the following:

- Private technology laboratories are classical industrial organizations that first populated and have long dominated (in numbers) the U.S. R&D laboratory community. It is likely that it is from this base that industrial labs have evolved to take on their newer operating environments.

Table 4.5	**Laboratory Distribution by Taxonomic Cell and Type**

Level of Market Influence	Level of Government Influence		
	Low	**Moderate**	**High**
Low	**Private Niche Science**	**Hybrid Science**	**Public Science**
		Total	Total
		Government 2	Government 52
	ISD	Industry 6	Industry 10
		University 33	University 63
		Other 13	Other 8
Moderate	**Private Science and Technology**	**Hybrid Science and Technology**	**Public Science and Technology**
	Total	Total	Total
	Government 0	Government 0	Government 37
	Industry 53	Industry 53	Industry 22
	University 3	University 13	University 14
	Other 3	Other 7	Other 4
High	**Private Technology**	**Hybrid Technology**	**Public Technology**
	Total	Total	Total
	Government 0	Government 1	Government 5
	Industry 170	Industry 76	Industry 17
	University 4	University 4	University 3
	Other 1	Other 1	Other 0

ISD = Insufficient Data

- Government laboratories are concentrated in the public science and the public science and technology environments, as might be expected. However, these two environments have fewer than 50% government-owned laboratories in their total population.
- Public technology laboratories are diverse in their sector characteristics indicating that government funds a variety of organizations. These include large numbers of industrial and university laboratories that are focused on commercial objectives.
- Industrially owned research organizations dominate the institutional framework of the U.S. research enterprise. However, these organizations are diverse in their operating environments.
- University laboratories are not exclusively dedicated to public domain scientific research. Over 35% of all university laboratories classified are significantly or moderately influenced by the market.

• Laboratory concentration by type indicates that most R&D facilities operate in simple environments with little conflict between public and market forces.

LABORATORY SIZE

By sector, the median number of scientific and technical employees in responding R&D laboratories is as follows: government 962, industry 282, university 194, and other (hybrid, multiform, or nonprofit) 320. In terms of the Environmental Context Taxonomy, the public technology laboratories are the largest research organizations (by number of full-time equivalent personnel), with an average of 820 professional personnel, and the hybrid science and technology labs are the smallest organizations, averaging fewer than 160 total professional personnel. In the case of the private technology laboratories, there are some large organizations, but most are small.

Average annual budgets range from $50–100 million in the public technology laboratories to the much smaller figure of less than $3–4 million in hybrid science laboratories. Again, laboratories with high levels of "publicness" are the largest and most heavily financed research organizations. Market-influenced laboratories that are not financed by the government are quite small, with an average annual budget of less than $8 million, excepting of course the large corporate labs.

RESEARCH MISSION

Research mission questions invariably interest policy makers. What laboratories are involved in basic versus applied research? What laboratories are involved in technology transfer activities? We have already indicated that expectations relating to research missions are often dashed since universities engage in commercially focused R&D and industrial laboratories produce public domain knowledge. Hybrids present a special challenge for understanding the research mission. Consider the Microelectronics Center of North Carolina (now simply MCNC). The laboratory is classified as a hybrid science and technology one and operates in a complex environment of multiple influences. More than 85% of the laboratory community in this group, including MCNC, view applied research as an important mission.

Not long ago, few individuals would have predicted that a development-focused microelectronics laboratory would be built by a state government in coordination with research universities. Now there is an entire class of such technology laboratories with substantial funding from both industry and government. These same types of labs have been developed by the federal government.

Lab Window 4.4: "Tobacco, Hogs & Microelectronics: Transforming the Tarheel State"

MCNC

http://www.mcnc.org/

MCNC is a unique corporation that offers cost-effective access to advanced electronic and information technologies and services for businesses, for state and federal government agencies, and for North Carolina's education communities to provide our clients with a competitive advantage. MCNC

1. maintains advanced capabilities—staff and facilities—in emerging electronic and information technologies. Partnering with a variety of customers, MCNC uses its resources to develop and apply technologies with commercial value, to help businesses integrate the latest innovations into their products, and to help build a competent workforce;

2. features two service centers for the state's universities. One runs a statewide network, providing internet access, data sharing, and video conferencing. The second center features a variety of world-class, advanced supercomputing resources;

3. leverages a history of leading-edge collaboration with research and development communities to work now with industry in many ways—through joint ventures, partnerships, and spin-offs.

MCNC's electronics technologists work with clients to develop and apply innovative solutions to microfabrication-based problems, working closely with businesses, government agencies, and researchers to reduce the time it takes to develop and commercialize new products and to improve our clients' competitive positions in the global marketplace.

MCNC's information technologists provide their customers with the networking and computing resources needed to compete internationally—partnering with a wide range of business, government, and academic customers, offering interoperability testing for new products and processes and applying new technologies for rapid product commercialization.

Of particular importance since the late 1980s has been the national interest in applied research. In recent years, numerous government policies have been developed to encourage more applied research in all laboratories. Among all U.S. R&D laboratories, applied research is identified as a major mission by more than 60% (at least 80% of all types of labs, except public science laboratories). Laboratories with high levels of market influence are virtually unanimous in their identification of applied research as a basic mission.

A vast majority of all research organizations operating in the U.S. today view applied research as a major reason for their existence. By contrast, basic research, which is the hallmark of the U.S. technical enterprise, is a major mission in far fewer laboratories than applied research. As shown in table 4.6, fewer than 10.5% of laboratories with high levels of market influence

Table 4.6	Percentage of Knowledge and Technology Output Devoted to Patents		

| Level of Market Influence | Level of Government Influence | | |
	Low	Moderate	High
Low	Private Niche Science ISD	Hybrid Science 1.7	Public Science 1.3
Moderate	Private Science and Technology 5.9	Hybrid Science and Technology 4.8	Public Science and Technology 3.2
High	Private Technology 6.5	Hybrid Technology 4.4	Public Technology 1.6

ISD = Insufficient Data

consider basic research to be a major mission, regardless of their level of government influence.

Among the laboratories with moderate levels of market influence, between 20% and 32% of the respondents identify basic research as a major mission of their research organization. In stark contrast, more than 62% of all respondents with low levels of market influence make the same basic research identification.

Size appears to be an important factor in the small number of highly market-influenced laboratories with basic research missions. Facilities such as Bell Laboratories (Lucent Corporation, formerly AT&T), the Intel Corporation, and the Exxon Research Center have the resources and organizational culture to maintain directed basic research missions, at least for now, even in the midst of serious market pressures. Of course, this is a dynamic environment with changes occurring often in the big hybrid science and technology labs. These organizations seem to be on a trend toward increased market-relevant research but nonetheless maintain the only major directed basic research agenda in industry.

Public science and hybrid science laboratories are the backbone of U.S. basic research. Basic research is not pervasive in any other laboratory types. In the U.S. basic research is highly concentrated, subject to strong influence from both the government and selected influence from the market, and is not a major function of most laboratories.

KNOWLEDGE AND TECHNOLOGY OUTPUTS

The R&D laboratory system's products interest policy makers as much as particular industrial labs' products interest CEOs. Companies and nations must attend to the composition of R&D. Recently, much debate has centered on the amount of applied research being carried out in the U.S. Concern has also been focused on whether or not there is an inappropriate emphasis on particular knowledge products such as patents or academic papers. There has also been on ongoing and intensive debate about the impact of increased industrial funding for academic research.

These concerns have many origins, a significant one being the desire of many policy makers to influence the knowledge outputs of various elements in the R&D laboratory community. Our NCRDP team focused significant attention on R&D output.

Tables 4.7 and 4.8 depict the range of knowledge outputs of U.S. laboratories. To a large extent, this distribution reflects traditional sector-based assumptions, even though there is a substantial ownership diversity in the various classification types. For instance, there is heavy concentration on the development of prototype materials and devices among all laboratory types with high levels of market influence.

In contrast, the laboratories with low levels of market influence and varied levels of governmental influence concentrate on the production of scientific and technical articles and conference papers. A brief review of the laboratory types and their characteristic outputs follows.

Table 4.7	**Three Primary Knowledge and Technology Outputs (average percentage of total output)**		
Level of Market Influence	**Level of Government Influence**		
	Low	**Moderate**	**High**
Low	**Private Niche Science**	**Hybrid Science**	**Public Science**
	ISD	Articles 43	Articles 46
		Conference 21	Conference 16
		Internal Reports 12	Internal Reports 13
Moderate	**Private Science and Technology**	**Hybrid Science and Technology**	**Public Science and Technology**
	Other 26	Prototype 22	Articles 27
	Prototype 25	Internal Reports 19	Prototype 16
	Internal Reports 18	Articles 16	Conference 12
High	**Private Technology**	**Hybrid Technology**	**Public Technology**
	Prototype 36	Prototype 37	Prototype 27
	Internal Reports 23	Internal Reports 20	Other 17
	Other 15	Other 13	Algrth Software 16

ISD = Insufficient Data

Table 4.8	**Laboratories' Scientific and Technical Output and Means by Sector and Taxonomic Category**					
Ownership	**Prototypes**	**Demonst.**	**Patent**	**Reports**	**Paper**	**Algorithms**
Government	10.3	8.5	2.1	16.7	12.9	8.2
Industry	32.6	9.6	5.3	20.0	4.4	8.2
University	6.3	4.0	2.1	10.5	18.8	7.0
Other	5.8	3.0	2.6	18.5	13.5	5.0
Taxonomic type						
Hybrid science	4.6	1.2	1.7	12.1	21.0	5.0
Public science	5.9	4.7	1.4	13.1	16.4	7.3
Private science and technology	25.0	7.6	5.9	17.1	5.2	7.7
Hybrid science and technology	22.3	9.6	4.9	19.1	7.6	10.7
Public science and technology	16.5	11.4	3.2	12.3	12.4	10.5
Private technology	36.0	8.8	6.2	23.6	3.5	4.7
Hybrid technology	37.1	9.6	4.5	19.6	3.7	9.6
Public technology	26.7	13.3	1.6	16.0	5.1	16.4

Hybrid Science Labs

A well-known example of a hybrid science lab is the Woods Hole Oceanographic Institution. This research laboratory is funded largely by the National Science Foundation, the Office of Naval Research, and various other agencies, foundations, and companies. It generates primarily public domain knowledge products. The government is important to the operation and facility base of this lab, giving it a quasi-public character. At the same time the lab works with private groups in industries as varied as fishing, insurance, energy, scientific instrumentation, and salvage to support their scientific needs. Research results can be of mixed character but, for the most part, they are generic and include scientific articles in peer-reviewed journals, technical reports, and conference proceedings.

Among hybrid science labs, 64% of the knowledge and technology outputs are of public character, usually in the form of reports and papers. Also included in this laboratory class are several, but not all, agricultural experiment stations that continue their traditional mission of academic research but with multiple funding sources. Examples include the stations in Minnesota, Nebraska, and Missouri. The Minnesota Experiment Station, for example, receives substantial funding from foundations and industrial groups with an interest in generic research products.

Lab Window 4.5: "Selling Agricultural Research in Minnesota"

Minnesota Agricultural Experiment Station
http://www.mes.umn.edu/~maes/

The Minnesota Agricultural Experiment Station is an important part of the University of Minnesota. It sponsors scientific research across a broad spectrum of interests. It also helps disseminate the results of that research, so the citizens of Minnesota can benefit from it.

For the past 100 years, Minnesotans have invested in their and future generations' welfare by funding research of the Minnesota Agricultural Experiment Station.

Established by the state legislature in 1885, the station organizes and supports research on the production, processing, marketing, distribution, and quality of food and other agricultural products. Other station research seeks to improve forests and forest products, public policy, human nutrition, family and community life, recreation and tourism, and overall environmental quality in Minnesota.

Faculty from the University of Minnesota's Colleges of Agriculture, Forestry, Home Economics, Veterinary Medicine, and Biological Sciences conduct most of the agricultural experiment station's research. The research is conducted on the university's St. Paul campus and at Rosemount, at branch experiment stations at Crookston, Grand Rapids, Lamberton, Morris, and Waseca and at other University of Minnesota research facilities as well as and on private and state-owned lands throughout Minnesota.

The people of Minnesota provide, in the form of appropriations from the state legislature, almost 62% of the resources for station research. Federal funds make up another 13%, and income from the sale of by-products of the station's animal and crop research provides 7%. Gifts and grants account for the remaining 18%.

A Justified Investment

Research is a gamble because it involves a probing into the unknown. Sometimes it results in no payoff or benefits. Sometimes it is of value only because it shows that something does not work. Sometimes the payoff from research is modest. At other times, however, the returns are large—even phenomenal, worth many times the original investment. In some instances, the benefits of research, although apparent, cannot be measured in dollars.

Have the investments made into the work of the Minnesota Agricultural Experiment Station been justified? The answer is a definite yes, whether the return is measured in money farmers make from improved crop varieties developed by station scientists or in such priceless terms as the improved quality of human lives and the better use of our world's finite resources. For example, the increased yield from just one variety of wheat developed in 1970 by the experiment station has yielded the equivalent of more than 6 billion additional one-pound loaves of bread, compared to varieties grown before; the development of particleboard created a whole new industry for Minnesota and the world; and diseases such as tuberculosis and brucellosis, which were common and sometimes fatal among cattle in the early years of this century, and which were communicable to people, have been eradicated from U.S. herds through the use of diagnostic tests developed by the experiment station.

Research That Benefits Us All

Although Minnesota farmers were the original beneficiaries of experiment station research, the mission of the station has broadened throughout its existence. Now all Minnesotans—and others around the nation and the world—share the benefits of station research, whether they live on farms or in cities, suburbs, or small towns. Basic research conducted by station scientists forms the foundation for the applied research that makes it easier, safer, more profitable, and more pleasant to work, live, and play in Minnesota.

What has the state received in return for the money it has invested in the Minnesota Agricultural Experiment Station through its more than 100 year history? Research supported by the station has paid handsome dividends to farmers, helping to keep agriculture "No. 1" among Minnesota's industries and assuring high-quality, reasonably priced food for all Minnesota consumers. Station research has helped Minnesota's agriculturally related industries to grow and remain competitive. Research by Minnesota Agricultural Experiment Station scientists has helped Minnesota's forest industries to grow and remain productive, and it has helped protect and improve the state's environment. And station research has done much to improve the quality of life for Minnesota Families.

Public Science Labs

These laboratories are similar to the hybrid science labs in that they have an almost exclusive concentration on public domain output. Public science labs are represented by some of the better known federal laboratories in the country including Goddard Space Flight Center, Fermi National Accelerator Laboratory, Lawrence Berkeley Laboratory, many of the NIH institutes, and all of the astronomy and astrophysics centers. These facilities are funded almost totally by the government and produce the classic public domain mix of "academic" papers and reports.

Lab Window 4.6: "Earth Science and the Space Systems that Enable It"

Goddard Space Flight Center
http://www.gsfc.nasa.gov/

HISTORY

Goddard Space Flight Center was created on 15 January 1959, named in commemoration of Dr. Robert H. Goddard, the American pioneer in rocket research. The first 157 employees were from the Vanguard project, transferred from the Naval Research Laboratory, Washington, D.C.

The mission of Goddard Space Flight Center is to expand knowledge of the Earth and its environment, the solar system and the universe through observations from space. To assure that the nation maintains leadership in this endeavor, the center is committed to excellence in scientific investigation, in the development and operation of space systems, and in the advancement of essential technologies.

MISSION

Goddard's role as a leader in technology and science is as alive today as it was in 1959 when Explorer VI, under Goddard project management, provided the world with its first image of Earth from space.

Goddard is the lead center in NASA's Mission to Planet Earth (MTPE) program. Mission to Planet Earth is NASA's long-term, coordinated research effort to study the earth as a global environmental system. The Earth Observing System (EOS) is the centerpiece of MTPE and is managed by Goddard. EOS will feature a series of polar orbiting and low inclination satellites for global observations of the land surface, biosphere, solid Earth, atmospheres, and oceans. The launch of the first EOS satellite, EOS AM1, is scheduled for 1998. The end product of MTPE will be the ability to develop and implement environmental policies based on a better understanding of how our environment works. To develop that understanding, MTPE will rely on the EOS Data and Information System (EOSDIS). The EOSDIS has been designed to archive, manage, and distribute MTPE data worldwide.

Approximately 12,450 persons work at the Goddard Space Flight Center at all of its sites. This number includes 3,517 civil servants and 8,932 contract personnel (Greenbelt, Maryland, 3,187 civil servants and 8,172 contract personnel; Wallops Island, Virginia, 312 civil servants and 729 contract personnel; Goddard Institute for Space Studies in New York, New York, 18 civil servants and 31 contract personnel). Of this number, 3,245 civil servants and 7,574 contract personnel reside in the state of Maryland. There are 32 major buildings, providing approximately 3,000,000 square feet (278,810 square meters) of space, located in Greenbelt, Maryland, situated on approximately 1,200 acres (485 hectares).

Major Contractors at Goddard, Greenbelt, Maryland, by Millions of Dollars Obligated

AlliedSignal Technical Services: 194.2

McDonnell Douglas Corp.: 165.4

Computer Sciences Corp.: 152.5

Santa Barbara Research Corp.: 93.5

Hughes Information Technical Corp.: 87.1

Lockheed Missiles & Space Co.: 81.0

Space Systems Loral, Inc.: 63.4

Hughes STX Corp.: 47.6

TRW Inc.: 41.3

Hughes Aircraft Company: 40.8

Ball Corp.: 39.0

Raytheon Service Co.: 34.5

Swales & Associates, Inc.: 33.3

Aerojet General Corp.: 32.2

Jackson & Tull, Inc.: 24.6

Manhattan Construction Co.: 24.6

NSI Technology Services Corp.: 23.8

Unisys Corp.: 18.7

ITT Corp.: 18.0

General Sciences Corp.: 14.5

Fairchild Space & Defense Corp.: 14.3

EER Systems Corp.: 14.0

Lawrence Berkeley Laboratory (LBL) offers a good illustration of the larger scale public science laboratory. Located near the campus of the University of California at Berkeley and operated by the UC system for the DOE, this laboratory focuses on the discovery of new knowledge and materials in areas related to future energy technologies. Staffed by some UC faculty and graduate students and by others from this same cultural mold, LBL is to a large extent a dedicated research university serving the public good in areas such as fusion energy and earth sciences. By design and despite its size (more than 3,000 employees), this laboratory represents organized academic science at its most targeted and focused. Simple sector-based classification would not provide an accurate picture of this laboratory.

Lab Window 4.7: "The Size of a Small Town"

Lawrence Berkeley National Laboratory
http://www.lbl.gov

HISTORY
Ernest Orlando Lawrence founded this lab, the oldest of the national laboratories, in 1931. Lawrence invented the cyclotron, which led to a golden age of particle physics and revolutionary discoveries about the nature of the universe. Known as a Mecca of particle physics, Berkeley Lab long ago broadened its focus. Today, LBNL is a multiprogram lab where research in advanced materials, biosciences, energy efficiency, detectors, and accelerators focuses on national needs in technology and the environment. Berkeley Lab is located above the UC Berkeley campus on land owned by the University of California system. Although not part of UC Berkeley, several hundred scientists have joint appointments both as lab staff and as UCB professors. Hundreds of UCB students work at the lab.

HISTORIC ACCOMPLISHMENTS
Of nine Nobel Prizes, five are in physics and four in chemistry.

STAFFING
Today, LBNL has some 3,400 employees, of which about 600 are students. Each year, the lab also hosts more than 2,000 participating guests.

OWNERSHIP
DOE, managed by the University of California.

In addition to the discovery and distribution of fundamental new knowledge, many public science laboratories focus on research and technical support missions that are intended to aid in the maintenance and development of the generic or infratechnology base. An example is the Stanford Linear Accelerator. This DOE lab, operated by Stanford University for the government, is a large-scale user facility for high-energy physicists from around the country. Its research results add to the science base, and the lab itself is a part of the national infrastructure for the conduct of research.

Lab Window 4.8: "The Search for Q: Examples of SLAC Research"

Stanford Linear Accelerator
http://www.slac.stanford.edu/

The Millicharged Particle Search (mQ) is a new experiment to search for unconventional, millicharged elementary particles. A millicharged elementary particle would have an electric charge three or more factors of ten smaller than the charge on the electron. We do not have a deep understanding of charge quantization; as far as we now know, theory does not exclude the existence of millicharged particles. Experimental bounds, which come from precise QED checks and high-energy beam-dump experiments, exclude $q/e > 10^{**}-2$ over a wide range of masses.

Very small charges ($q/e < 10^{**}-6$) are disfavored by astrophysical arguments, but particles with $10^{**}-6 < q/e < 10^{**}-3$ could exist in nature. Theorists have suggested that millicharged elementary particles may constitute the so-called dark matter or account for other experimental anomalies. This experiment will run parasitically during routine SLC operation, which requires production of positron beams.

Electroproduction is the mechanism for producing millicharged particles, and is fully calculable, allowing a definitive search over a wide range of charge and mass. Unlike other beam-dump experiments, this search will exploit a low-energy signature in the ionization and/or excitation in the atoms by using a scintillation detector to search for very small energy depositions. Understanding the scintillation mechanism at the single-photon level is an important component of the experiment.

This experiment presents an opportunity often missing from large-scale high-energy physics experiments: participation in all aspects of an experiment, from detector construction, to Feynman diagram evaluation, to experiment operation, to data analysis.(1)

Group K is a particle astrophysics group at the Stanford Linear Accelerator Center (SLAC). The research activities of the group lie in the area of X-ray and gamma-ray astrophysics. Current projects involve collaboration on the Unconventional Stellar Aspect (USA) satellite experiment, analysis of the archival X-ray data from HEAO A-1 and BATSE data on gamma-ray bursts. For the longer term, there is an effort on a Gamma-ray Large Area Silicon Telescope (GLAST) that could lead to a follow-on space mission to the Energetic Gamma-ray Experiment Telescope (EGRET).

The Stanford Synchrotron Radiation Laboratory, a division of SLAC, is a national user facility that provides synchrotron radiation, a name given to X-rays or light produced by electrons circulating in a storage ring at nearly the speed of light. These extremely bright X-rays can be used to investigate objects of atomic and molecular size, allowing the over 1,000 researchers from 170 institutions who use the SSRL facility each year to perform basic and applied studies on the structure of matter. The facility, which provides 26 experimental stations and ancillary equipment, is used by researchers from industry, government laboratories, and universities in many areas, including the fields of biology, chemistry, geology, materials science, electrical engineering, chemical engineering, physics, astronomy, and medicine.

Private Science and Technology Labs

Because it is independent of government influence and is not completely driven by the market in terms of its research agenda, the Philips Petroleum Laboratory is representative of the private science and technology laboratory type. While most of the knowledge produced is kept within the corporate laboratory structure, some specialized and more public outputs are produced. These are targeted to specific narrow science areas of interest to the firm and are similar to the science outputs of the public science labs but without external consumption. As a result, research is conducted both with purely scientific as well as with market goals in mind. Thus it appears that this type of lab is willing to fund and carry out some research for which the payoff may be minimal, indirect, or long-term in character.

Lab Window 4.9: "From Uncle Frank to 3-D Seismic Technology"

Phillips Petroleum Laboratory
http://www.phillips66.com/

Phillips is the only major oil company named for its founder. Frank Phillips, known by employees as "Uncle Frank," was a remarkable businessman.

Uncle Frank's first well struck oil but quickly fizzled out. Second and third wells were dry holes. The desperate Phillips brothers barely had enough money to drill a fourth well on a lease obtained from a young Delaware Indian named Anna Anderson. Their luck changed stupendously: The Anna Anderson No. 1 came in a gusher, the first of a fantastic string of 82 consecutive producing wells.

The Phillips oil businesses did well for several years. But then—as now—oil prices followed boom-bust cycles. In 1916 prices hit bottom, and Uncle Frank decided to sell out and stick to banking. However, prices soared when the United States entered World War I in 1917. Uncle Frank decided to stay in oil and consolidate operations in one firm, known as the Phillips Petroleum Company. It was incorporated in Delaware on 13 June 1917.

Uncle Frank guided the new business with vision. Phillips Petroleum pioneered the production of natural gas at a time when competitors saw gas as a nuisance. It was one of the first companies to pursue extraction and sale of such natural gas liquids as propane and butane. In the 1920s Uncle Frank put the firm in a leadership role in petroleum research, then moved it into refining and marketing.

Phillips' research placed it in a strategic role during World War II because of breakthroughs in high octane aviation gasoline and artificial rubber.

A Leader in 3-D Seismic Technology
Oil companies thump the ground and then listen to the echoes bouncing back from rock layers beneath the surface. This seismic process helps them learn more about the subterranean geology and identify spots where oil deposits are likely to be found.

For years, it has been next to impossible to locate oil deposits underneath layers of salt because salt severely distorts traditional seismic data and makes interpretation almost impossible.

Phillips geoscientists and engineers proved otherwise 80 miles offshore in the Gulf of Mexico. In an area many thought was depleted, Phillips used supercomputers plus new, sophisticated visualization and processing techniques to correct the distortion and generate a clearer, more accurate image of the rock beneath the salt.

The result? Discovery of the Mahogany oil field and confirmation of an exciting new technology.

Phillips has led the industry in 3-D seismic imaging and interpretation techniques that help improve the exploration process and lower the cost of finding and development.

A key difference between this laboratory type and the more typical private technology lab is that some of the knowledge/technology output of these laboratories is made available for general consumption by the science community. A number of special knowledge/technology products, such as developed systems, make up the largest average percentage of output. In addition, prototype outputs account for an average 25% of the total output makeup of this laboratory type.

Hybrid Science and Technology Labs

Perhaps the best known laboratory of this genre is Bell Laboratories of Lucent Technologies, Inc., which we have aggregated into a single "lab" in spite of its several sites. As a result of public and private research investment in both facilities and project funding, the organization is a hybrid. However, it is owned by a corporation and is devoted to the production of public scientific knowledge as well as commercial or market-oriented products. The knowledge outputs of this type of R&D organization benefit the entire technical enterprise, in terms of both generic and market breakthroughs resulting from combined basic and applied research efforts.

In general, hybrid science and technology laboratories manifest an academic air. Research is organized in university-type groups with many investigators free to pursue their individual research agendas. In this arrangement, fundamental discoveries are common and expected. Bell Labs has historically been known throughout the world for its ability to sustain fundamental discovery and to capitalize on new inventions.

There is a wide mix of both public and private knowledge and technology outputs in this laboratory type. The largest output is in new prototype materials, processes, and devices, followed by internal reports, but articles for scientific publication are also common. The combined market and public influences seem to provide for a broad mix of scientific outputs.

Other laboratories in this type include GTE, Alcoa Laboratories, and The Battelle Memorial Institute. In the case of Battelle, both the funding base and the output base are diverse. Some research groups at Battelle generate new prototype materials and processes as a prerequisite to survival, while other groups conduct basic science research with no particular payoff.

Lab Window 4.10: "A Not-for-Profit Making For-Profits Profitable With New Technologies"

Battelle Memorial Research Institute
http://www.battelle.org/

Battelle has delivered technology-based value to industry and government for more than 65 years. Battelle develops the technology behind the products of some of the most successful companies in the world. Its focus is on developing high-quality products and reducing time-to-market for our clients. The institute inserts technology into systems and processes to turn problems into opportunities for manufacturers, trade associations, pharmaceutical and agrochemical industries, and government agencies supporting energy, the environment, health, national security, and transportation. A staff of over 8,000 technical, management, and support professionals meet the commercialization needs of clients in more than 30 countries.

Commercial and Industrial Technology
Battelle got its start because an industrialist wanted technology to deliver practical solutions to commercial and industrial problems. The institute thrived during the Great Depression, the Second World War, and in a constantly changing marketplace because "the winners" in industry and government discovered that innovative technology—harnessed to strategic goals—is a great competitive advantage.

Competitive advantage counts now more than ever. Global competition demands even more rapid and effective product introduction and improved manufacturing. Success requires technological innovation and effective integration of evolving technologies into business solutions.

Battelle realizes commercial value from innovative technology—faster, more reliably, and more fully than anyone. And that translates into competitive advantage.

Battelle is your hidden asset—using technology to enhance and accelerate product development, refining processes to increase efficiency and respond to environmental requirements, and helping you strategically manage technology to stay ahead of the competition.

Lab Window 4.11: "Telephones Are Not Low-Tech"

GTE Laboratory
http://info.gte.com/

GTE, with annual revenues and sales exceeding $21 billion, is one of the largest publicly held telecommunications companies in the world. It also is the largest U.S.-based local telephone company and a leading cellular-service provider—with wireline and wireless operations in markets encompassing about a third of the country's population. Outside the United States, where GTE has operated for more than 40 years, the company serves over 6 million wireline and wireless customers. GTE is also a leader in government and defense communications systems and equipment, aircraft-passenger telecommunications, directories and telecommunication-based information services and systems.

GTE Laboratories traces its roots back to 1943, to the establishment of the Sylvania Technical Center, the research and development center for Sylvania Electric Products, Inc. When General Telephone and Electronics Corporation (now GTE) purchased Sylvania in 1959, the Sylvania Technical Center became the nucleus of what is now GTE Laboratories. In 1972, GTE Labs moved from Bayside, Long Island, to its current location in Waltham, Massachusetts.

Today GTE Laboratories, under the direction of C. David Decker, president, undertakes research and development for the full range of technologies required by GTE business units. Work performed at GTE Labs supports ongoing technical needs and assures GTE Corporation that adequate technical options will be available in the future for the company's core telecommunications businesses.

To accomplish this, GTE Laboratories is organized into six major organizational units: the Advanced Systems Laboratory, the Network Infrastructure Laboratory, the Operations Systems Laboratory, the Services Research Laboratory, Software Systems Laboratory, and the Administration group.

Public Science and Technology Labs

These organizations are best characterized as government-financed laboratories with prototype devices, materials, and processes for development as major products, particularly when compared with public science laboratories. The FAA Technical Center in New Jersey and the Naval Research Laboratory represent the government version of these multiproduct laboratories.

Scientists at the Naval Research Laboratory, for instance, have focused both on fundamental discoveries and the development of such technological products as new sonar configurations. For example, the NRL has recently continued a long tradition of developing new powder metallurgy techniques that could aid in the production of advanced permanent magnets. In this instance, the Navy's interest is to capitalize on these fundamental material discoveries and then develop a prototype demonstrating the production process.

As a result, industries supporting the Navy could then use or produce such magnets with greater ease. Thus, at the NRL, one finds a Nobel laureate in science as well as someone who is schooled in the needs of industry and production matters together in a lab owned and operated by the government.

Lab Window 4.12: "Those Ain't Your Grandpa's Torpedoes"

Naval Research Lab
http://www.nrl.navy.mil

The Naval Research Laboratory (NRL) is the Navy's corporate research and development laboratory, created in 1923 by Congress for the Department of the Navy on the advice of Thomas Edison. The laboratory has over 4,000 personnel (over 1,500 full-time scientists, engineers and SES employees—more than half of these Ph.D.'s, currently including a Nobel laureate), who address basic research issues concerning the Navy's environment of sea, sky, and space. Investigations have ranged widely from monitoring the sun's behavior, to analyzing marine atmospheric conditions, to measuring parameters of the deep oceans, to exploring the outermost regions of space. Detection and communication capabilities have benefited from research that has exploited new portions of the electromagnetic spectrum, extended ranges to outer space, and provided means of transferring information reliably and securely, even through massive jamming. Submarine habitability, lubricants, shipbuilding, aircraft materials, and fire fighting along with the study of sound in the sea and the advancement of radar technology have been steadfast concerns. New and emerging areas include the study of biological and chemical processes and nanoelectronics.

MISSION
To conduct a broadly based multidisciplinary program of scientific research and advanced technological development directed toward maritime applications of new and improved materials, techniques, equipment, systems, and ocean, atmospheric, and space sciences and related technologies.

The Naval Research Laboratory provides
1. primary in-house research for the physical, engineering, space, and environmental sciences
2. broadly based exploratory and advanced development programs in response to identified and anticipated Navy needs
3. broad multidisciplinary support to the Naval Warfare Centers
4. space and space systems technology development and support

STAFFING
At the end of FY 93, NRL employed 4,079 personnel—42 officers, 74 enlisted, and 3,963 civilians. Among the research staff, there are 924 employees with doctorate degrees, 493 with

master's degrees, and 707 with bachelor's degrees. The support staff assists the research staff by providing administrative, computer-aided designing, machining, fabrication, electronic construction, publication, personnel development, information retrieval, large mainframe computer support, and contracting and supply management services.

Private Technology Labs

Private technology labs, the traditional industrial labs, focus almost exclusively on the technical goals of the industrial parent or sponsor. Illustrative laboratories in our study include Pfizer Central Research, Texaco Research Center, Eastman Kodak's Chemical Division Laboratory, and Lederle Labs, to mention just a few of the larger ones. Most important is the fact that, for the most part, this class is made up of small and somewhat isolated (in organizational terms) laboratories. These are typically small engineering-oriented shops with a narrow product line and a narrow research and technology development agenda.

Lab Window 4.13: "DeepStar: Cooperatively Leveraging Texaco Research"

Texaco
http://www.texaco.com/deepstar/deepstar.htm

WHAT IS DEEPSTAR?

The DeepStar Project identifies and develops economically viable, low-risk methods to produce hydrocarbons from deep water tracts in the Gulf of Mexico. The U.S. Gulf contains significant reserves and remains one of the best domestic oil company development opportunities, but producing hydrocarbons in water this deep presents commercial and technical challenges.

Here's where DeepStar comes in. DeepStar is the industry's best mechanism for adapting to this new market. Participants COOPERATE to develop the technology necessary to tackle deep water. In doing so, each participant minimizes the cost and risk of developing the technology while at the same time making the most of its particular technology achievements.

WHO SPONSORS DEEPSTAR?

Agip, AMOCO, ARCO, BHP, British Gas, BP, Chevron, Conoco, Elf, Exxon, Kerr McGee, Marathon, Mobil, OXY, PETROBRAS, Phillips, Shell, and TEXACO.

Plus 34 manufacturing and service companies—click on the sunrise and check them out!

DeepStar provides a working interface between industry and government agencies to address regulatory issues and critical technology development issues. It is supported by the MMS of the U.S. Dept. of the Interior.

DeepStar supports the research of university labs, including the Offshore Technology Research Center at Texas A&M.

Lab Window 4.14: "Keeping the Pipeline Full"

Pfizer Pharmaceuticals
http://www.pfizer.com/rd/rdfrm.html

Pfizer's success in marketing new, differentiated, and value-added products flows from its commitment to research and development. Taking the form of aggressive investments in facilities, in equipment, and in the expansion of an enthusiastic, highly productive staff, the commitment has resulted in a pipeline of new products unparalleled in our history.

The Birthplace of Tomorrow

Central Research, employing more than 4,000 professionals worldwide, is where it all begins. An attractive 90-acre campus in Groton, Connecticut, is the home of our R&D headquarters, with other R&D sites located at Sandwich, England; Nagoya, Japan; Terre Haute, Indiana; and Amboise, France. In this global network of laboratories, Pfizer scientists are currently developing a broad range of new product candidates to treat a variety of human and animal diseases.

The New Age of Gene Discovery

For several years, drug research has been undergoing a period of unprecedented opportunity for therapeutic innovation. Powerful tools of molecular genetics and recombinant DNA technology have given Pfizer access to isolated human genes. This development has revolutionized the drug discovery process. More than half of the discovery approaches in Pfizer's current portfolio are based on genetic technology.

Access to novel genes has become a determinative factor in Pfizer's ability to initiate new discovery approaches in its pharmaceutical, food science, and animal health businesses. Teams of biologists, chemists, physicians, statisticians, toxicologists, veterinarians, engineers, and countless other professionals with B.S., M.S., and Ph.D. degrees pour their knowledge, their experience, and their passions into the search for new therapies, confident that Pfizer has the resources and expertise to successfully bring new products to market.

Central Research's findings are vital and diverse. Discoveries literally run the gamut from A to Z: Advocin, a potent, new antibiotic used to treat respiratory diseases in livestock; Norvasc, a new calcium channel blocker to treat cardiovascular diseases; and Zoloft, a new once-a-day antidepressant.

Fully 60% of all outputs in private technology laboratories are of a propriety character, with 36% devoted to new prototype materials and devices. These laboratories also provide "other" products, most of which are designed to serve the parent companies and the laboratories' clients.

On the larger side, take for instance the Eastman Kodak Chemical Division (now Eastman Chemical). For many decades, this company has produced specialty chemicals as an adjunct to its film-making business. Some of these chemicals are needed in photography; others are easy to produce while the processes are underway. As a result, Kodak had a small, at least in terms of total corporate sales, business center in these chemicals. Supporting this effort through the years has been Kodak's Chemical Division Research Laboratory in Rochester, New York.

This lab now focuses technically on selling intermediate chemical goods to industrial customers. Their research is evolutionary and for the sole benefit of the company. Typical projects focus on improving the production costs of existing products and producing a new, slightly different, product for a particular consumer.

Lab Window 4.15: "Spinning Off Corporate Central Research"

Eastman Chemical Company
http://www.eastman.com

Eastman Chemical Company (NYSE: EMN) is a leading international chemical company that produces more than 400 chemicals, fibers, and plastics. The company does not sell consumer products, but Eastman supplies billions of pounds of products to industrial customers for use in the manufacture of hundreds of items consumers use every day. Eastman polyethylene terephthalate (PET) plastic is used in containers for beverages, foods, and toiletries. Several brands of tools, toys, toothbrushes, and eyeglasses feature products made from other Eastman plastics. Eastman products are in fabrics, floor coverings, paints and other coatings, computers, automobiles, pharmaceuticals, foods, cigarette filters, and chemical intermediates.

Founded in 1920 as a unit of Eastman Kodak Company, the small operation in Kingsport, Tenn., was established to produce two basic chemicals for Kodak's photographic business. After 73 years as a Kodak unit, Eastman Chemical Company was spun off on 1 January 1994 and is now an independent public company whose shares are traded on the New York Stock Exchange. While Kodak is Eastman's largest single customer, Eastman has more than 7,000 customers worldwide. Eastman's sales in 1996 totaled $4.8 billion.

Headquartered in Kingsport, Eastman employs about 17,500 people in more than 30 countries. The company has regional headquarters in The Hague, Netherlands; Miami; Kingsport; and Singapore. Principal manufacturing facilities are located in Tennessee, Texas, South Carolina, Arkansas, New York, West Virginia, Canada, Spain, Mexico, Hong Kong, England, and Wales.

Although there is a much higher level of public support and influence in mixed-source technology laboratories, the output categories for this type are almost identical to those of private technology. The main difference is that these laboratories are heavily involved in government contract research or government procurement support. Examples include Magnavox Advanced Products Lab, and the Allison Turbine Division Lab of General Motors.

Lab Window 4.16: "MAGNAVOX: A History of Smart Innovations"

Magnavox Advanced Product Lab

http://www.magnavox.com/companyinfo/corpoverview/historyofmag.html

For the past 75 years, Magnavox innovation and leadership have consistently been recognized as the industry hallmarks of reliability, state-of-the-art technology, and consumer acceptance.

Begun in 1911 in a small shop in Napa, California, with a total area of 800 square feet and total sales in its first year of operation of exactly $477, the Commercial Wireless & Development Company, as it was originally named, was founded by Peter Jensen and E. S. Pridham, with Richard O'Connor as their principal financial backer with an investment of $2,500, which was soon to pay big dividends and open the door to newfangled instruments of which the public had never dreamed.

One such instrument was the world's first electrodynamic telephone, invented in that same year, 1911; it was an invention of such importance that it laid the foundation for present advances in the art of sound amplification.

Four years later, breakthroughs in the field of sound reproduction became a byword at the young company. For example, a true industry milestone was reached with the invention of the first electrodynamic loudspeaker, a speaker that is today on permanent display at the Greenfield Village Museum in Michigan. For many years after that, Magnavox was the only manufacturer of loudspeakers in the United States.

In 1915 the company officially adopted the name Magnavox. Derived from the Latin words "magna" and "vox," and meaning "great voice," it was chosen because of the "great voice" produced by Magnavox loudspeakers.

Also in 1915 music was amplified for the first time, with the Magnavox invention of the first amplifier phonograph. This instrument, drawing its power from a 6-volt storage battery, was the forerunner of today's sophisticated and popular component music system.

The Magnavox milestones of 1915 didn't stop there. Also introduced in that year was the first public address system, an innovation that made well-deserved headlines across the country, because for the first time, the size of the audience was not limited by actual "hearing distance" from the speaker.

Ex-president William Howard Taft used this first public address system in speaking to a large gathering at Grant Park in Chicago. Woodrow Wilson became the first United States president to use a Magnavox public address instrument during a speech heard by more than 75,000 people in San Diego, California.

1911 - First electrodynamic telephone	1952 - 27" Television
1915 - Dynamic loudspeaker	1957 - Portable short-wave radio
1915 - First amplified phonograph	1958 - First hi-fidelity stereo
1922 - First amplified radio phonograph	1960 - Jam-proof record player
1937 - Hi-fi phonograph	1961 - Videomatic - black & white TV
1949 - Black & White television	1962 - Solid-state stereo

1964 - Automatic fine-tuning color TV

1967 - 82-channel VHF/UHF tuning

1969 - Total automatic color

1971 - Complete home entertainment line

1974 - Star television - first total electronic tuning system

1974 - Odyssey - first home video game

1976 - Touch-tune television

1978 - Laser video disc

1978 - Computer color

1978 - First use of high-resolution filter

1979 - Television with hi-fi audio

1983 - Laser audio disc

1986 - Universal remote control

1986 - Videowriter word processor

1987 - Compact disc - video

1988 - Magnavox "Smart Window" (picture-in-picture)

1992 - Magnavox "Smart Sound" (automatic volume control)

1993 - Magna bass

1993 - Remote locator

Lab Window 4.17: "The Progress of Engines, the Engines of Progress"

Allison Turbine Division Labs
http://www.allison.com/

MISSION

Allison Engine Company designs, manufactures, markets, and supports gas turbine engines and components for aviation, marine, and industrial applications with over 140,000 engines produced as of April 1995. Located in Indianapolis, Indiana, U.S.A., Allison employs approximately 4,300 people at its research, engineering, and manufacturing facilities. This includes the Single Crystal Operation, also located in Indianapolis, which is dedicated to the development and fabrication of advanced single crystal engine parts and components for Allison Engine Company and other customers; the AERO Repair and Overhaul facility, located in Indianapolis and dedicated to developing engine overhaul and repair methods supporting field operators; and the Allison Evansville Operation, located in Evansville, Indiana, and dedicated to the development and fabrication of advanced sheet metal products for engines and exhaust systems.

BRIEF HISTORY

1915 - Founded by James A. Allison as the Indianapolis Speedway Team Company

1929 - Purchased by General Motors becoming the Allison Division

1932 - Delivery of the first V1710 engine to the U.S. Navy

1955 - Delivery of the first T56 for the C-130 aircraft

1963 - Received contract to build components for Apollo space program

1970 - Allison merged with Detroit Diesel to form Detroit Diesel Allison Division

1983 - Gas turbine operations separated from Detroit Diesel Allison as Allison Gas Turbine

1992, April 2 - GM announces the intent to seek a buyer for Allison Gas Turbine Division

1993, December 1 - Allison Gas Turbine became Allison Engine Company

1994, November 21 - Agreement to sell Allison Engine Company to Rolls-Royce plc announced

1995, March 24 - Rolls-Royce plc receives government approval and completes acquisition of Allison. Allison Engine Company becomes a member of the Rolls-Royce Aerospace Group

Facilities and Workforce

4 Facilities located in Indianapolis, IN, totaling 3,737,000 square feet

1 Facility located in Evansville, IN, totaling 56,000 square feet

approx. 4,300 total employees

Products

Large Engines - T56/501D Series III, T56/501D Series IV, T406-AD-400, AE 2100, AE 3007

Small Engines - Model 250-C18 Series I, Model 250-C20 Series II, Model 250-C30 Series IV, T800-LHT-800

Industrial Engines - 501K, 570K, 571K, AG9130 Navy Gensets

In most of these cases, the government strongly supports the development and final "commercialization" of a needed technical product by funding the corporate lab at various stages of its research. For instance, if the Air Force is interested in continued production of advanced electronics and communication devices, it is in its best interest to be involved in the various stages of research.

As a result, in some labs there is tremendous government R&D support for development-oriented research as well as DOD independent research and development support for intermediate-range applied and development research.

Public Technology Labs

Public technology laboratories are basically government-sponsored, industrial research institutes (Crow, Emmert, and Jacobson, 1991). As such, they operate with a public support base and an industrial technology focus. Public technology labs include two major varieties. First, there are the large, industrially owned laboratories in the defense and space industries that are funded almost completely by the government. Since the government is the only market for high-performance jet fighter components and advanced electronic countermeasure technologies, large, government-financed labs in the "private sector" are common. The same is true for certain areas in the field of aerospace and space technology. Examples include laboratories such as the historic North American Aircraft Laboratory (now a part of Rockwell International) and the Space and Systems Technology Group of Motorola.

Lab Window 4.18 "From Beepers to National Defense"

Motorola Space and Systems Technology Group
http://www.mot.com./GSS/SSTG

Motorola's Space and Systems Technology Group specializes in the research, development, and production of advanced electronic systems and equipment for the U.S. Department of Defense, NASA and other government agencies, commercial users, and international customers. Undoubtedly, one of the most significant contributions to global communications technology is our involvement with the IRIDIUMR global personal communications system. The project, once operational, will permit voice, data, fax, and paging services from anywhere to anywhere else on the planet with a wireless, hand-held device. We have been contracted to design, construct, and launch this revolutionary system.

Business Units

Communications System Division (CSD)

Information Security Division (ISD)

Diversified Technologies Division (DTD)

Communications Test Equipment (CTE)

Diversified Technology Services

Radio Systems Operation (RSO)

Tactical Systems Operations (TSO)

Satellite Communications Group (SATCOM)

Advanced Systems Division (ASD)

Government Space Systems Division (GSSD)

Ground Systems Division (GSD)

Mobile Satellite Systems Division (MSSD)

The second major variety of public technology lab consists of the industrial service labs established by various governments to assist industries with a low natural rate of R&D investment or to fill a market failure gap. These labs address specific industrial problems with public support. One example is the Kentucky Center for Applied Energy Research, which has focused on building and supporting the coal and synthetic fuels industry in Kentucky. Other examples include the various former research stations of the U.S. Bureau of Mines, such as the one at Pittsburgh (closed in 1996), which has been involved in all five phases of the NCRDP. These stations served the U.S. mining industry partly because of the industry's own lack of research activity, particularly with respect to safety concerns.

Lab Window 4.19 "A University Non-Academic Research Unit"

Center for Applied Energy Research
http://www.caer.uky.edu

The Kentucky energy research program began over twenty years ago as farsighted leaders in government, industry, and academia aggressively responded to the immense opportunities offered by unprecedented national commitment to development of technologies to produce fuels, especially transportation fuels, from coal. The center's programs have evolved as federal energy policy has changed. Today, while the Center for Applied Energy Research (CAER) continues to focus on research relating to coal-derived synfuels in recognition of the certainty that the U.S. will turn to synfuels to ensure the nation's energy security, it is increasingly concentrating on challenges and opportunities in an era of utility industry deregulation and escalating environmental demands. Over the last twenty years, energy research at the CAER has brought many benefits to the Commonwealth. Some of these are as concrete as increased productivity, new uses for coal, decreased costs of using coal, and disposing of or utilizing coal combustion wastes. Some, such as leadership in research and in national coal policy both now and in the future, are somewhat intangible but are nonetheless very important.

The Center for Applied Energy Research (CAER) is one of the University of Kentucky's 20 multidisciplinary research centers and institutes within the Research and Graduate Studies sector. It is located in a stand-alone facility approximately 11 miles from campus. The CAER is a nonacademic unit that is staffed by professional scientists and engineers, has extensive inter-actions with faculty, and provides support and facilities for faculty research and student research and employment.

The center's research provides a focal point for energy and environmental research at UK and in the Commonwealth of Kentucky. Research programs focus on energy resources indige-nous to Kentucky with efforts directed to advanced beneficiation, utilization, conversion, and process technologies. Environmental issues relating to fuel use and by-product wastes consti-tute a growing efforts, along with the derivation of high-added-value materials and chemicals from energy resources.

Additionally, CAER serves as an information and sales agency for IEA Coal Research pub-lications. The IEA Coal Research Center is a collaborative project established in 1975 and involving member countries of the International Energy Agency (IEA).

Public technology laboratories are heavily financed and influenced by the government, but they have an output profile similar to that of labs in the private technology category. They serve their industrial clients with the pro-duction of prototype materials and devices as well as with problem-solving algorithms and software that the clients are unable to produce or finance effectively for themselves.

Implications of Classification Typology for R&D Laboratories and Public Policy

To understand the U.S. R&D laboratory community and to plan the focus, outputs, and system roles of the federal labs, we must revise the operating model. The environmental context model provides an alternative conceptual model, one separating laboratories according to their mix of environmental influences. This classification permits the analysis of the environmental influences, and the typology has identified labs of similar character and behavior. Within these groupings, variations in laboratory roles and functions are captured.

Public science and private technology laboratories are the dominant organizational forms, perhaps accounting for as much as one half of the total number of R&D laboratories in the U.S. system. These laboratories also appear to be the "pure" types from which most variations have evolved over the decades. That is not to say that the private technology laboratory of the 1990s is the same as the industrial lab of the 1890s. But the 1990s private technology lab remains focused on research problems related only to the business lines of the parent corporation.

Within the public science category, the traditional government lab of 1930, with its technical service mission and government employee base, still exists. Today these labs focus on energy, agriculture, weights and measures, and environment. But today's public science laboratories also include new entrants, including industrial and university labs. The focus of these labs is as diverse as science itself.

Beyond the archetypal public (public science) and private (private technology) form, the U.S. R&D laboratory is made up of 19% hybrid science or public science and technology labs, both of which have many characteristics similar to those in the private technology category. In both cases, there are evident differences in market and governmental influences, but the laboratories themselves are similar. Thus, an additional 40% of the U.S. research system is made up of labs that have a dominant public or private influence as well as a secondary influence.

The remaining public technology and hybrid science and technology laboratories differ greatly from the traditional industrial and governmental laboratories in their character, mission, output, and other traits. While they are the largest set of producers of broad knowledge outputs, they are overlooked in most assessments of the research arena.

EFFECT OF GOVERNMENT INFLUENCE ON R&D LABORATORIES

Our data indicate that over 65% of all U.S. R&D laboratories experience at least a moderate level of direct governmental influence, with more than a third being heavily affected by the government. NCRDP studies show that government influence is associated with the following behaviors and characteristics of R&D laboratories (Bozeman and Crow, 1995; Crow and Bozeman, 1991):

- More basic research
- More cooperative research
- More bureaucratization and red tape
- Faster release of new knowledge
- More technology transfer to the commercial sector
- Heavier emphasis on technology transfer to the government
- Moderate to high levels of applied research
- Increased focus on scientific effectiveness
- Heavy dependence on government funding and thus an inability to pursue other funding
- Stability for enhanced R&D productivity
- Outmoded research equipment
- Greater and more numerous barriers to R&D productivity
- Generally larger research organizations
- Higher levels of interorganizational complexity
- Mixed knowledge outputs including both proprietary and nonproprietary products

EFFECTS OF MARKET INFLUENCE ON R&D LABORATORIES

Increased public influence on R&D laboratories has not come at the expense of market influence but, usually, is built upon market influence. Market influence is also on the rise in many laboratories. Thus the overall complexity of R&D laboratories' environments is dramatically increasing. Because of policy shifts many government-owned labs, particularly those conducting public science, have been asked to increase their market relevance. Within the R&D laboratory system, market influences are substantial within more than 70% of all R&D laboratories (Bozeman and Crow, 1995). Market influence is significant in almost 30% of all university-owned laboratories and in 44% of all government-owned ones. Market influences on R&D laboratories generally have the following effects (Bozeman and Crow, 1995; Crow and Bozeman, 1991):

- Strong focus on applied research
- Lower levels of cooperative research
- Slower release time for scientific and technical knowledge

- Concentration in engineering and chemistry
- Less interdisciplinary research
- Smaller size (except for public market and hybrid science and technology laboratories)

ENVIRONMENTAL CONTEXT

The Environmental Context Taxonomy has significant limitations, particularly in the absence of time series data. Nevertheless, we can provide some data-based speculation. In particular, we feel that public science and private technology lab types may be the "genotype" from which "mutations" occur. But patterns of environmental selection require additional data for their verification.

Our cross-sectional snapshots indicate that the distribution of R&D is subject to three driving forces. First, there are the technology-policy and technology-driven laboratories: public science, hybrid science and technology, and public science and technology laboratories. These groups contain most of what would be called the national labs and large private sector labs, such as Bell or IBM, that are players in the NIS.

Lab Window 4.20 "IBM Global Research Portfolio"

IBM

http://www.research.ibm.com/research/intro.html

IBM Research has approximately 2,600 employees, with more than half of its researchers holding Ph.D.'s. It encompasses seven laboratories across the world:

The Thomas J. Watson Research Center, founded in 1961 in Westchester County, New York, is the largest research facility, where roughly half of the division's employees work. Research in this laboratory focuses on semiconductors and both physical and computer sciences as well as mathematics.

The Almaden Research Center in San Jose, California, dedicated in 1986, with approximately 700 employees, focuses on storage systems and technology, computer science, and science and technology. It is the successor to the San Jose Research Laboratory, which was established in 1952.

The Austin Research Laboratory, established in 1995 in Austin, Texas, focuses on advanced circuit design, as well as new design techniques and tools, for very high performance microprocessors. The Austin Lab currently consists of approximately 30 researchers.

The China Research Laboratory opened in Beijing in September [1996] and employs approximately 20 researchers. It will be the site of specialized research focusing on Chinese language and speech recognition as well as on digital library technology research and applications in China.

The Haifa Research Laboratory in Haifa, Israel, was formed in 1982. About 230 scientists and engineers conduct research into applied mathematics, computer science, and engineering.

The Tokyo Research Laboratory founded in 1982 in Yamato, Japan, consists of approximately 250 people and specializes in computer science, storage and semiconductor technology, and manufacturing research.

The Zurich Research Laboratory, located in Rueschlikon, Switzerland, with approximately 200 employees, was established in 1956. Areas of concentration include communication systems and related information technology solutions, optoelectronics, and physical sciences.

A second grouping includes market-driven laboratories: private technology, private science and technology, and hybrid science and technology laboratories. Finally, there are the public technology laboratories, the most complex of the types, as they are pushed and pulled by policy, technology, and the market.

These differences among laboratory types have important consequences for public policy in terms of funding and policy instruments and also in the allocation of responsibilities among sectors and laboratory types. As new approaches and concepts for classifying laboratories continue to develop, whether as simple as the notion of a Federally Financed Research and Development center (FFRDC) or as complex as the environmental context typology we have proposed, the need for public policies sensitive to institutional nuances is enhanced.

The Environmental Context Taxonomy's implications for federal laboratories are particularly striking. We conclude:

1. Federal laboratories are diverse in mission and environment. The result is that broad-based policies insensitive to diversity have little prospect of success.

2. Federal laboratories are already deployed across a wide spectrum of activity. These activities are intended to address an equally diverse set of missions. The result of wide mission dispersion is that there is no likely simple mission adjustment for any federal laboratory.

3. Instability is inherent in federal labs with high levels of government control and high levels of market influence.

4. Federal labs resemble university labs or industrial labs with similar foci more than they resemble other federal labs.

5. Many federal labs are highly acculturated in academic culture.

Recommended Use of the Environmental Context Taxonomy

The Environmental Context Taxonomy is not an analytical tool like benefit-cost analysis or program evaluation. The taxonomy is akin to the standard industrial classification codes (SIC) used by the Department of Commerce to help us understand the kinds of business entities operating in the U.S. The principal difference is that we have gone beyond simple descriptions of outputs like the ones the SIC system uses and have gone on to a classification mechanism that separates organizations according to functions, culture, and operating environment. The taxonomy is a sorting tool.

What this means for science and technology policy is that policy making needs to consider the class or type of R&D organization when considering policy options. For instance if in the general policy-making arena there is a desire to see an increased U.S. capability in advanced computing, policy makers should focus their resources and policy assets on the class of R&D organizations with the means to be responsive. If the U.S. government continues on the path of supporting national prominence in the area of basic science, policies specific to math and engineering and directed toward the public science laboratory community should be pursued.

As we will illustrate in later sections, policy making now as it relates to the R&D laboratory community is poor. There are assumptions about lab behavior and potential that have allowed the "national labs" of the Department of Energy to become a political football in the policy debate about what a national lab should be. As we will argue later it is what it is and can't be anything else. R&D laboratories are functions of their environment. Change their environment, and you obviously change them. Change them, and you change the knowledge production and technology production capability dramatically. Change this capacity, and you change the basis on which the lab operates, as opposed to going to a lab for what it does well and building on that base.

When national policy makers think about our defense assets they don't think of the Boeing Company as a great aerospace company that could also make tanks or submarines. They think about it for what it is and try to shape an industrial policy accordingly. When the government thinks about the overall national innovation system it obviously doesn't think about it in an industrial policy sense. But the government should think about it in terms of assessing the national assets that are needed to maintain national technological quality. In this context, therefore, one can easily imagine that the use of this taxonomy could be used to ask and deal with questions such as those detailed below.

BUILDING THE SCIENCE BASE FOR THE NATIONAL INNOVATION SYSTEM

- What is the general health and performance of the nation's public science labs? How have these labs been performing and what is the relative quality of their knowledge products?
- How do the public science labs and the public science and technology labs interact with academia in general and what mechanisms for enhancement might be considered?

- What mechanisms can be developed to enhance the linkages between public science labs, public science and technology labs, and the private technology lab community—specifically what can be done to speed knowledge transfer without adversely effecting the performance or culture of either community?
- Between the public science labs and the public science and technology labs, which organizations are best equipped to take the lead in certain subjects and fields?

FEDERAL INVESTMENT IN SCIENCE AND TECHNOLOGY

- What is the effectiveness, measured in technical, financial, and social terms, of the various federal R&D investments in each of the various laboratory types?
- What has been the impact of the Small Business Innovation Research Program in the birth and death rate of private technology labs and hybrid technology labs? Why the difference?
- Where (in what laboratory types) should the federal government concentrate its investments in research in terms of maintaining a national capability for basic research?

NATIONAL TECHNOLOGY PROJECTS

Is there adequate public technology and private technology laboratory capacity to respond to national technology project needs as they arise?

National Technology Development Programs

The National Institute for Standards and Technology of the U.S. Department of Commerce has assumed that there is a need for a national funding program to catalyze private technology development in certain fields. What is the impact of such a program on laboratory performance and outputs, and more importantly, why are these funds needed or desired in the private technology sector? More specifically, what is it that the public technology labs or the hybrid technology labs are doing with their already diverse public/private influence environment that the private technology labs are not doing? Why would an agency implement such a program without being alert to such differences?

Lab Window 4.21 "NIST: We're Not Just Standards Anymore"

National Institute of Standards and Technology
http://www.nist.gov

NIST's four major programs are designed to help U.S. companies achieve their own success, each one providing appropriate assistance or incentives to overcoming obstacles that can undermine industrial competitiveness.

Advanced Technology Program (ATP) — provides cost-shared awards to companies and company-led consortia for competitively selected projects to develop high-risk, enabling technologies during the preproduct phases of research and development.

Manufacturing Extension Partnership (MEP) — a growing nationwide network of extension centers, cofunded by state and local governments, that offers small and medium-sized manufacturers access to technical assistance as they upgrade their operations to boost performance and competitiveness.

Laboratory Research and Services — develops and delivers measurement techniques, test methods, standards, and other types of infrastructural technologies and services that provide a common language needed by industry in all stages of commerce—research, development, production, and marketing.

Baldrige National Quality Program — with industry, manages the Malcolm Baldrige National Quality Award, which may be awarded annually to companies in the categories of small business, manufacturing, and service and provides U.S. industry with comprehensive— and extensively used—guides to quality improvement.

Staff
About 3,300 scientists, engineers, technicians, and support personnel, plus some 1,250 visiting researchers each year.

Budget
About $810 million, estimated FY 96 resources from all sources.

BASIC RESEARCH IN INDUSTRY AND BELL LABS AS A NATIONAL ASSET

Few would argue that the Bell Labs of the old monopolistic days of AT&T were not a unique national treasure. Their scientific and technological accomplishments between 1940 and 1980 were simply second to none. Given the changes with AT&T's breakup and the subsequent changes with the formation of Lucent Technologies as the Bell Labs' parent, what should government do, if anything, to maintain the national core competencies that may be resident in Bell Labs? Given that these labs have historically been hybrid science and technology labs with a complex influence matrix, what might the government do to insure their survival or influence their course? In this same context the hybrid science and technology laboratory has historically played a critical role in the NIS. Should the government strive to maintain excellence in this laboratory type?

We hope the Environmental Context Taxonomy renders the mysterious 16,000 U.S. R&D laboratories at least a bit less enigmatic. The recognition that much of the variety among the 16,000 results from the riptides produced by the convergence of market and political forces does not qualify as a great leap forward in understanding R&D laboratories. We are not the first to note the joint impacts of political and economic authority on R&D labs. But the fact that the application of a typology based on those two pillars of environmental change seems to distinguish laboratory behaviors, structures, and performance at least gives hope of an alternative to treating the U.S. R&D laboratory system as, essentially, a black box.

The Environmental Context Taxonomy conforms with the institutional design approach we advocate; indeed, it embodies the approach. The *systemic principle* makes clear the need for knowledge of R&D laboratory environments, and the taxonomy provides one approach to developing systemic knowledge. While there is nothing about the Environmental Context Taxonomy that yields particular policy prescriptions, it gives us a framework for prescriptions and a way of thinking strategically about sets of laboratories. We began this chapter by noting our doubt that the halls of Congress will soon be abuzz with taxonomic discussion. But we hope that policy professionals and policy analysts, persons devoting larger proportions of their time and energy to reflections about R&D laboratories, will find in the Environmental Context Taxonomy new analytical tools and modes for expressing ideas about R&D laboratories.

We provide a modest test of the Environmental Context Taxonomy but much remains to be examined. Those interested in laboratory-environment relations and their strategic implications have a full plate. First, is it possible to chart the "mechanics" of movement within the system? How does a laboratory move from one environmental context to another? What is the rate of movement within the system? Second, what are the prescriptive heuristics? Can one develop guidelines or rules of thumb about appropriate levels of concentration in particular environments? How do we know if we have a shortfall or a glut?

Third, what varieties do we find in particular environments? We have set to work on this last question. Chapter 5 employs the Environmental Context Taxonomy and focuses on federal laboratories. Given particular environmental types, as outlined in the taxonomy, what are the subtypes? We examine the variety in environments common to federal laboratories and apply the typology by providing public policy conclusions flowing from the analysis of environmental types.

CHAPTER FIVE

Federal Laboratories and the National Innovation System

Applying the Environmental Context Taxonomy

Systems-thinking about R&D laboratories, including federal laboratories, has been in short supply. But inattention to the whole does not imply neglect of the parts. Indeed, one set of important "parts," the federal laboratories, has for years received a remarkable degree of attention. Just a few decades after their immediate postwar status as heroic beyond reproach, the federal laboratories have, by this point, been poked and prodded and (sometimes literally) studied to death.

The basis for the continuing fascination with federal laboratories is less mysterious than the activities of the labs. First, they represent the single largest deployment of organized research units in the country, with over $25 billion in expenditures and over 130,000 employees. Second, federal labs result directly from a government policy formulation process and therefore are of keen continuing interest to policy makers and policy analysts. Third, federal labs are in greater flux than any set of labs in the country, their budgets are continually threatened, and their missions are unstable. Finally, many of these labs perform roles that no other lab or set of labs performs.

In the preceding chapter, we presented a broad classification typology for all labs; the Environmental Context Taxonomy. There, we took some pains to show that federal laboratories are distributed throughout the various performance niches of the NIS. In this chapter we take the typology and apply it, adding detail, to R&D labs owned or established by the United States government.

Viewing Federal Labs Through the New Typology

In this chapter we provide an enhanced framework for thinking about these labs within the context of the proposed new typology. The laboratory types presented as part of the environmental context typology can, in the analysis of federal labs, be broken further down to six subcategories (Crow and Bozeman, 1989b):

1. pure basic research laboratories,
2. directed basic research laboratories,
3. intermediate-range applied research laboratories,
4. technology development laboratories,
5. infratechnology laboratories, and
6. industrial service laboratories.

In the terms and categories of the Environmental Context Taxonomy, over 90% of all federally owned labs are either public science labs, public science and technology labs, or public technology labs. These laboratory types are considered below in terms of subtypes, or "varieties."

PUBLIC SCIENCE LABORATORIES

Focused on basic research and on the pursuit of fundamental knowledge public science labs represent about 20% of the R&D labs operating in the U.S. Of these, about 40% were established by the federal government. While nearly all receive their principal funding from the federal government, the fact that 40% were established by the federal government and serve the government makes these labs the single largest national asset available in the basic research arena. Within the public science laboratory type, federal labs perform two basic research functions, detailed below.

Pure Basic Research Labs

These federal laboratories, regardless of their formal mission, are dominated by a curiosity-driven science agenda. None of these federal laboratories pursues pure basic research exclusively; indeed, few laboratories focus exclusively on any technical mission. Usually, there is a balance among several missions. The pure basic research labs have basic research as their dominant orientation, not just as one of many activities on the research spectrum. As such, the basic research orientation affects the lab's culture, its formal and informal reward systems, and its approach to setting its research agenda. The knowledge products have public goods characteristics (i.e., knowledge freely available and communicated in such open forums as scientific journals).

One example of a pure basic research lab is the Kitt Peak National Observatory, which focuses on stellar mapping and astrophysics. Kitt Observatory has more than one set of technical activities, but exists chiefly to provide basic research.

Lab Window 5.1 "Where Staring Is Okay"

Kitt Peak Observatory, NSF
http://www.noao.edu/kpno/kpno.html

MISSION

NSF supports the National Optical Astronomy Observatories (NOAO), the national center for research in ground-based optical and infrared astronomy and solar physics. Large optical telescopes, observing instrumentation, and data analysis equipment are made available to qualified visiting scientists. The NOAO staff of astronomers, engineers, and various support personnel is available to assist visiting scientists in their use of the facilities.

The headquarters of NOAO are in Tucson, Arizona. NOAO includes the Kitt Peak National Observatory (KPNO), the Cerro Tololo Inter-American Observatory (CTIO), and the National Solar Observatory (NSO). NOAO is operated and managed by AURA. The observing facilities of KPNO are located on Kitt Peak, a 2,089-meter mountain 90 kilometers southwest of Tucson. The facilities include the 4-meter Mayall Telescope, a 2.1-meter general-purpose reflector, a 92-centimeter coudé feed (associated with the 2.1-meter), a 1.3-meter reflector instrumented for the infrared, and the Burrell Schmidt Telescope of Case Western University. A full complement of state-of-the-art spectroscopic, photometric, and imaging instrumentation is available for use on these telescopes.

In Kitt Peak, as in other federal government pure basic research labs, the reward system resembles academic units of universities: publications in refereed scientific journals and recognition among their peers are the "payoff" for individual researchers. Kitt Peak is to a large extent staffed by active university research personnel. As in other nations, the primary observatory and astronomy apparatus is government-owned. The flow of information among these nations' basic research astronomy laboratories is free and open with the scientists embracing the same universality norms reported in most basic research-oriented segments of the scientific community.

A second example of a pure basic research lab is the Stanford Linear Accelerator of the Department of Energy. Here, a group of physicists, including both full-time on-site researchers and academic researchers from around the world, focus on the ongoing quest to understand matter and energy at their most fundamental level. While the results of their research certainly have an impact throughout science and ultimately technology, the knowledge is pursued solely for its intrinsic worth. This "knowledge for its own sake" is the hallmark of a pure basic research laboratory.

Pure basic research labs contribute little to the stock of knowledge and technology put to near-term use by U.S. industry. Their economic benefits are measured less precisely and never easily (e.g., Bozeman and Rogers, in press; Bozeman and Kingsley, in press). The indicators of success of these laboratories include scientific awards, incidence and impact of knowledge in refereed scientific journals, and reputation within the appropriate worldwide scholarly communities. Other examples of this type of federal lab include several of the biomedical labs located on the campus of the National Institutes of Health in Bethesda, Maryland, elements of the Naval Research Lab in Washington, D.C., and various lab structures established with funding from the National Science Foundation.

In general, though, it must be noted that the number of organized labs (as opposed to the smaller faculty-led research groups in a university setting) focused on pure basic research is very small. The vast majority of basic research in the U.S. is carried out by university faculty funded by the federal government.

Directed Basic Research Labs

The line between directed basic research labs and pure basic research labs is in many ways a fine one, but the conceptual distinction can be made. Pure basic research labs are curiosity-driven without any significant consideration to mission, potential application, or general problem category. Directed basic research labs, while likewise concerned with fundamental knowledge, pursue knowledge with clear assumptions about payoff and possible long-term applications. To many scientists, such distinctions are trivial or wrongheaded. Most research scientists—whether federal employees, industrial researchers, or university professors—view basic research as any research where the pathway is uncertain. But to the policy analyst the key question in drawing a distinction between directed basic and pure basic research is, "why do you want to know?" If one wants to know the basic physics of plasmas to develop a better fusion energy system, one is doing directed basic research. If one plays with high-order math functions because they are fun and may perhaps prove someday useful to others in studying the basic physics of plasmas, then one is doing pure basic research. The fundamental question here is "why is the research being done?" If the answer is to solve a known problem related to an instrumental purpose, then the research is directed basic research. If the research is driven by nothing more specific than, say, learning how sun spots come about, then it is pure basic research. While this distinction means little to most scientists, it is a critically important one to policy makers. The majority of federal labs in the public science arena are devoted to a directed basic research agenda.

An example of a directed basic research lab is the Princeton Plasma Physics Lab (PPPL), now being downsized. Here, the basic research is aimed clearly at producing and controlling thermonuclear fusion. PPPL basic research focuses on understanding the physics of low-density plasmas produced and confined to magnetic fields. Despite the fact that economically relevant results are expected no sooner than in a 100 years or so, the existence of a mission framework colors the laboratory. For example, there is some expectation that research will occur on a trajectory and that progress will be made toward the broad mission goal. This affects planning processes and project selection, thereby distinguishing the directed basic research lab from the pure basic lab. Still, the similarities are greater than the differences. The two basic research lab types tend to have similar cultures, rewards, and performance measures. In each case, the outputs are academic papers and publications.

Lab Window 5.2 "Our Single Purpose Is/Was Fusion"

Princeton Plasma Physics Lab
http://www.pppl.gov

The Princeton Plasma Physics Laboratory (PPPL), located on Princeton University's James Forrestal Campus in Plainsboro, N.J., is a single-purpose fusion laboratory that is operated by Princeton University and funded by the U.S. Department of Energy (DOE). PPPL is a world leader in the development of magnetic fusion energy as a safe, economical, and environmentally acceptable method of generating electricity.

Magnetic fusion research at Princeton began in 1951 under the code name Project Matterhorn. Lyman Spitzer, Jr., professor of astronomy at Princeton University, had for many years been involved in the study of very hot rarefied gases in interstellar space. Inspired by the fascinating but highly exaggerated claims of fusion researchers in Argentina, Professor Spitzer conceived of a plasma being confined in a figure-eight-shaped tube by an externally generated magnetic field. He called his concept "the stellarator" (star generator) and presented this design to the Atomic Energy Commission in Washington. As a result of this meeting and a review of the invention by designated scientists throughout the nation, the stellarator proposal was funded and Princeton University's controlled fusion effort was born. In 1958 magnetic fusion research was declassified allowing all nations to share their results openly.

Current Status
At $225 million, the Department of Energy's proposed budget for magnetic fusion research in fiscal year 1998 is essentially the same as 1997. However, DOE has proposed $47.6 million for PPPL, a 15% cut from the current fusion budget of $55.9 million. This decrease follows a reduction of 14% in 1997 and 35% in 1996. The 1996 cut resulted in the loss of 240 jobs at PPPL, with approximately 500 remaining. The budget proposed for 1998 will result in a layoff of up to 200 individuals this spring. While some downsizing was expected as a result of the Tokamak fusion Test Reactor (TFTR) shutdown, the laboratory did not anticipate a reduction of this magnitude. The decrease in funding will impact the vitality of the National Spherical Torus Experiment (NSTX) program.

Despite the very severe cutbacks PPPL is determined to maintain a high-quality fusion program. Our strengths are very closely aligned with the goals of the restructured fusion sciences program. PPPL envisages a strong, diverse program of experimental and theoretical plasma science, emphasizing understanding of the physics of plasmas and innovation in fusion confinement concepts. In the 1999–2004 time frame, PPPL sees a vibrant NSTX experimental program, along with a second moderate-sized advanced toroidal device. PPPL will be working with DOE to make this vision a reality.

Public Science and Technology Laboratories

While only 10–11% of the nations' labs are of the public science and technology type, this is a critical lab type and federal labs represent a high percentage in it. These are labs that the government largely supports but with mixed influence from market forces. These labs have the capacity to do both fundamental science and technology development and, in the context of technology development, to work with a moderate level of market influence. This means that the knowledge and technology results have a more near-term purpose and a more near-term payoff. Nearly half of these labs are federal institutions, meaning that the federal government's stake in the production of new knowledge and technology with near-term implications is quite significant to these labs and their planning. Federal labs in the public science and technology lab category are devoted to applied research and to the development of basic infratechnology knowledge and systems. Each variety is reviewed below.

Intermediate-Range Applied Research Laboratories (IRARL)

This lab type is highly significant in that it bridges the gap between the two types of basic research labs and the commercially dominated laboratories of the federal lab system. There are three IRARL subvarieties.

The first subvariety of IRARL, as distinguished by foci, is the *large-scale research lab* such as Brookhaven or Argonne National Labs. While many view these labs as the centers for basic research they once were, they have evolved over the last several decades into centers focused on solving a wide range of scientific problems. The problem might be related to a new material and its development, cleaning up nuclear waste, or a new national technology venture such as the national battery initiative of the mid-1990s. In any event, these labs are focused on using large-scale facilities and advanced, highly capable research groups to attack large-scale and small-scale applied research problems. A good deal of pure or directed basic research occurs in these settings. But the preponderance of the research is carried out as a part of applied objectives of the lab and the Department of Energy. How is this different from directed basic research? The principal difference is timing. At the public science laboratory focused on directed basic research, the problem set may require decades for its resolution. At the public science and technology lab, the actual work is similar but the solution set focuses on a horizon of 5–10 years, and the work has specific objectives.

Lab Window 5.3 "The Nation's First National Lab"

Argonne National Laboratory
http://www.anl.gov

Argonne National Laboratory is one of the U.S. Department of Energy's largest energy research centers, with an annual operating budget of about $470 million supporting more than 200 research projects. Argonne was the nation's first national laboratory. Today, the laboratory has about 4,500 employees, including about 1,700 scientists and engineers, of whom about 850 hold doctorate degrees.

Argonne research falls into four broad categories:

- The Advanced Photon Source will soon provide the nation's most brilliant X-ray beams for pioneering research in materials science, a cornerstone of technological competitiveness.

- Energy and environmental science and technology includes research in biology, alternative energy systems, environmental assessments, economic impact assessments, and urban technology development.

- Engineering research focuses on advanced batteries and fuel cells, and advanced fission reactor systems—including electrochemical treatment of spent DOE fuel for disposal, improved safety of Soviet-designed reactors and technology for decontaminating and decommissioning aging reactors.

- Physical research includes materials science, physics, chemistry, high-energy physics, mathematics, and computer science, including high-performance computing and massively parallel computers.

Industrial technology development is an important activity in moving benefits of Argonne's publicly funded research to industry to help strengthen the nation's technology base. This activity at Argonne grows out of basic and applied research in materials, transportation, computing, advanced manufacturing, energy, the environment, and other areas. The laboratory maintains an electronically searchable database of programs and capabilities in which partnerships with industry are invited. Argonne transfers technology to industry through

1. collaborative research and development agreements (CRADAs),
2. licensing,
3. work for others, and
4. "quick-response" mechanisms.

Since 1984, Argonne has been the source of more than 30 spin-off companies.

Another important part of Argonne's mission is to design, build, and operate national user facilities. These facilities, such as the Advanced Photon Source and the Intense Pulsed Neutron Source, are large, sophisticated research facilities that would be too expensive for a single company or university to build and operate. They are used by scientists from Argonne, industry, colleges and universities and other national laboratories, and often by scientists from other nations.

Argonne's Division of Educational Programs provides a wide range of educational opportunities for faculty and students ranging from leading national universities to local high schools. More people attend educational programs at Argonne than at any other DOE national laboratory.

Argonne is operated by the University of Chicago for the U.S. Department of Energy.

Consider recent statements by the recently retired director of the Argonne National Laboratory, with 4,500 employees—another lab of this variety. On 17 June 1996, in a speech before a group in Chicago, Alan Schriesheim stated, "My laboratory, Argonne, developed and tested an inherently safe nuclear reactor—one that emits no air pollution, produces little waste, consumes waste from other nuclear plants" (Schriesheim, 1996). This is clearly long-range applied research and development backed up by basic research. Presently, major Argonne programs include high-density battery development for electric vehicles and a laser welding system for advanced steel car manufacturing. In addition, Argonne operates a number of major user facilities focused on enabling basic research. Facilities such as the Intense Pulsed Neutron Source are used to rapidly jump from fundamental knowledge production to practical solutions.

The second IRARL subvariety is the *basic-to-applied bridge laboratory*. This type of laboratory is often found in federal regulatory agencies. Examples include the National Exposure Research Lab of the U.S. Environmental Protection Agency at Cincinnati. This lab conducts intermediate-range applied research, anticipating clear applications but not in the near term, and focuses chiefly on the area of pollution control device technologies. The lab strives to bridge the gap between basic research and currently feasible technology, usually through applied research associated with targeted development programs.

Lab Window 5.4 "Have You Been Exposed?"

Human Exposure Research Division
NERL, ORD, U.S. EPA
http://www.epa.gov/nerlcwww/herd.htm

The Human Exposure Research Division of the National Exposure Research Laboratory (NERL) operates within the Office of Research and Development (ORD) and conducts research to measure, characterize, and predict the exposure of humans to chemical and microbial hazards. This research will provide information on environmental pathways on which hazardous contaminants are transported via air, water, food, and soil to populations at risk. Analytical quantitative methods are developed to accurately and specifically measure human risk factors associated with inhalation, ingestion, and dermal pathways. Surveys and monitoring studies are carried out to determine the levels of hazardous chemicals and microbials in environmental matrices, and human populations are studied to determine significant exposure pathways, the levels of exposure, and the sources of exposure factors. State-of-the-art analytical methods are used to measure organic and inorganic chemicals. Genomic and immune-based methods, as well as traditional cultural methods, are used to measure hazardous bacteria, viruses, fungi, and protozoa. Molecular- and sero-epidemiological tools are used to assess human populations for evidence of exposure to environmental hazards. The division conducts its multidisciplinary research program with a broad skill mix of scientists that includes organic, inorganic, and analytical chemists, bacteriologists, virologists, parasitologists, immunologists, and molecular biologists. The division consists of three branches, the Biohazard Assessment Research Branch, the Aerosol and Aquatic Exposure Research Branch, and the Chemical Exposure Research Branch.

A third subvariety of IRARL is the *codevelopment bridge laboratory*. This type is illustrated by such Department of Defense laboratories as the Army Construction Engineering Lab and the Army Cold Regions Research and Engineering Lab. Laboratories of this type focus their efforts on intermediate-range applied research so that industry can codevelop new products needed by the Department of Defense. Technical activities involve both applied research and demonstrations.

Lab Window 5.5 "The Army Doesn't Stop Just Because It's Cold"

Army Cold Regions Research and Engineering Lab
http://www.crrel.usace.army.mil/

The Cold Regions Research and Engineering Lab (CRREL) is a unique facility of the U.S. Army Corps of Engineers that addresses the problems and opportunities unique to the world's cold regions. CRREL has earned an international reputation of excellence that is sustained by its exceptional technical and support staffs and by emphasis on a balance of theoretical, experimental, laboratory, and field work.

Below 32°F or 0°C, the natural world changes dramatically, and snow, ice, and frozen ground become the dominant conditions. Most materials change their properties, and many machines either don't work as designed or fail completely. CRREL exists largely to solve the technical problems that develop in cold regions, especially problems related to construction, transport, and military operations. Most of these difficulties arise because water changes into ice. In other words, they are caused by such things as falling and blowing snow, snow on the ground, ice in the air and in the ground, ice in rivers, ice on seas and lakes, and ice in man-made materials.

Nearly half the earth's surface is subject to snow, ice, and seasonally frozen ground. As much as 20% is on top of permafrost, and 10% of the oceans can be covered with ice. Over 80% of Alaska contains permafrost, and the Arctic Ocean, an area of nearly 5.5 million square miles, is almost completely covered with ice year-round. These cold regions of the world are not only of great strategic importance, but also hold a major share of the earth's natural resources.

The mission of the Cold Regions Research and Engineering Laboratory is to gain knowledge of the cold regions through scientific and engineering research—and to put that knowledge to work for the Corps of Engineers, the Army, the Department of Defense, and the nation. CRREL is the DOD's only laboratory that addresses the problems and opportunities unique to the world's cold regions.

Cold regions research is based largely on the earth sciences (including atmospheric and ocean sciences) and on the basic physical sciences. Knowledge gained from more fundamental studies is developed and applied through the various branches of engineering and technology. Together with its predecessor organizations, the Snow, Ice and Permafrost Research Establishment (SIPRE) and the Arctic Construction and Frost Effects Laboratory (ACFEL), CRREL has been in business for about 40 years, and has been involved in all of the major U.S. cold regions ventures during that period.

CRREL serves the Corps of Engineers and its clients in three main areas:

1. traditional military engineering, which deals with problems that arise during conflict;
2. military construction and operations technology, i.e., the building and maintenance of military bases, airfields, roads, ports, and other facilities; and
3. civil works, which involves the corps in such things as navigation on inland waterways, coastal engineering, and so forth.

CRREL also deals with the problems of cold regions for the other defense services, for civilian agencies of the federal government, and to some extent for state agencies, municipalities, and private industry.

Overall, IRARL variety of labs is unusually complex in its technical activities, objectives, evaluation processes, and environmental influences. The mix of knowledge production and technology production influences, like the mix of market constraints and political constraints, usually implies more difficulty in laboratory management and greater instability.

Infratechnology Labs

The Infratechnology Labs in the NIS are almost all federal. The chief product of infratechnology labs is not new scientific research or technology development (in any traditional sense), but rather knowledge and technology supportive of the overall U.S. scientific and technical enterprise. Infratechnology labs have been prominent in other governments' laboratory systems. Japan, in particular, has created a number of infratechnology labs aimed at providing standards and coordinating structures for technology development support prior to the emergence of a dominant design from industry (Papadakis et al., 1995; Bozeman and Pandey, 1994).

In the U.S. the laboratories of the National Institute for Standards and Technology are perhaps the most familiar, but these are not the only examples. Within the NIST structure, the Information Technology Lab (ITL) is an excellent example of an infratechnology laboratory. With more than $30 billion in annual R&D expenditures in the information technology industry (Morella, 1996), the ITL lab in Maryland focuses on the development of industry-wide measurement issues and the development of national reference standards. These standards permit the U.S. industry to compete more effectively in international markets and help in many ways to secure international standards for electronic devices and materials that are at least compatible with U.S. interests.

Much less publicized, the National Seed Storage Laboratory of the USDA, in Ft. Collins, Colorado, collects and analyzes a wide range of seed samples. The purpose of this effort is to maintain a national storehouse of all the "seed-embodied" technology developed naturally and through plant breeding. In this way, the genetic purity of particular plant varieties can be preserved for later use. There is no private sector incentive to store seed technology that does not have a viable product associated with it in the near term.

The laboratory of the U.S. Geological Survey in Reston, Virginia, is an infratechnology lab. At this lab, data are developed on the extent of resources and reserve availability of all geological resources, including water. While there are private sector incentives to accumulate knowledge about reserves, the collection effort would be highly fragmented in the private sector and would focus on resources with currently known commercial applications.

PUBLIC TECHNOLOGY LABORATORIES

One of the lessons we learned in setting up the typology discussed in chapter 4 is that the more complex the influences, the more difficult, unstable, and unmanageable the operating environment. Public technology labs have high levels of governmental influence and high levels of market influence making this lab environment the most complex. Perhaps for this reason fewer than 4% of the labs in the nation are of this type. Of these, only 20% are in federal ownership. These federal labs are noted for having difficulties in operations. Within this group, the federal presence is limited to a few labs dedicated to technology development and a few that act as research labs for industrial sectors with poor or no internal capacity for research. Each of these subvarieties is reviewed below.

Technology Development Labs (TDL)

Among the most visible lab types, these facilities range from Los Alamos National Laboratory, historically charged with developing weapons technology such as nuclear devices, to the U.S. Coast Guard R&D laboratory (now in transition) in Groton, Connecticut, where until recently the focus was on the near-term development of new technologies for use by the Coast Guard. We chose these two examples to demonstrate that TDL labs are diverse with respect to size, resources, complexity of technology development objectives, and range of technical activities. What places these otherwise diverse labs in the same broad category is that each is trying to develop and produce technology for either a well-designed government mission or for a clear-cut set of select clients.

Another example of a TDL is the NASA Ames Research Center in California, which has been working to develop, among other things, new technologies for advanced robotic-supported, unmanned space vehicles. As a means of stimulating the development of this technology, NASA-Ames has distributed information and prototypes as widely as possible and hopes to entice private sector labs to contribute to the development of this technology. Since the market for this NASA-Ames technology is largely in the future, the lab's technology and market-creating activities go hand in hand.

Lab Window 5.6 "From Mining Research to Synthetic Fuels to Clean Coal"

Pittsburgh Energy Tech Center
http://www.petc.doe.gov/

The Federal Energy Technology Center in Pittsburgh, commonly referred to as FETC (PGH or Pittsburgh), is one of the federal government's principal fossil energy research centers, responsible for the technical and administrative management of fossil-energy-related research and development programs. Its origins date back to 1910 when Congress created the U.S. Bureau of Mines.

FETC has changed along with the country's needs, beginning with a mission related to basic coal research and evolving to research and development of synthetic fuel technology and technologies designed to use fossil fuels in a more efficient and environmentally acceptable manner.

FETC really began to take shape in 1944 with the passage of the Synthetic Liquid Fuels Act, which authorized the Bureau of Mines to build energy research laboratories called the Bruceton Research Center. As research into synthetic fuels continued, coal research efforts expanded over the next three decades.

Two key events in the 1970s—passage of the Clean Air Act in 1970 and the energy crisis generated by shortages—expanded the Bruceton Research Center's role. In 1970 the Center was renamed the Pittsburgh Energy Research Center, and research grew to keep pace with an increase in federal energy programs.

Among the new research assignments was the development of advanced methods for cleaning coal and combustion gases, along with alternative methods to substitute coal for imported oil. With the creation of the U.S. Department of Energy in 1977 came the creation of PETC.

PETC now is one of the federal government's principal fossil energy research laboratories, responsible for the technical and administrative management of coal-related research and development programs. PETC's major programs cover the entire cycle of coal use—from coal preparation to advanced combustion to flue gas cleanup. The emphasis of the work is on improving the environmental acceptability of the nation's vast supplies of coal.

PETC also oversees research on improved processes to use U.S. coal resources to produce liquid transportation fuels and chemicals. The center complements its contract management with an in-house program of fundamental, exploratory, and applied research. Using state-of-the-art equipment and expertise, PETC scientists and engineers not only conduct hands-on research but also provide technical expertise to evaluate and monitor R&D contracts with industry and universities.

PETC also maintains an integrated set of core competencies that include managing technical programs for the Office of Fossil Energy; managing over 600 active research and development contracts, grants, and cooperative agreements with industry and universities; providing direct support to U.S. industry to supply the global economy with energy and environmental products and services, creating value-added American jobs; managing the physical plant and personal property and environmental health and safety concerns for other DOE facilities; and providing administrative services to DOE elements such as procurement, finance, legal, human resources, planning, communications, and safeguards and security.

FETC also plans and implements 25 cooperative agreements between private-sector participants and the federal government under the Clean Coal Technology Program. This multiyear, multibillion-dollar, cost-shared partnership is designed to demonstrate a new generation of innovative coal technologies offering improved environmental benefits, higher efficiencies, and lower costs.

The Department of Energy's Pittsburgh Energy Technology Center (PETC) is another example of a TDL. In this laboratory the goal is the development and demonstration of new technologies for cleaning coal prior to combustion. Since it is federal policy to stimulate development of a variety of energy resources and, at the same time, ensure environmental quality, PETC is positioned to develop technologies that deal with these two somewhat conflicting goals. Pursuit of conflicting goals is common in a government TDL. Industry laboratories often have difficulty in developing technologies where there are conflicting goals, particularly when those goals are subject to somewhat unstable government regulations.

The output from technology development laboratories is often in the form of reports, internal working documents, and prototypes. If applied science for the public domain is produced in the form of scientific journal articles, it is, essentially, a secondary product of the laboratory. A key question is, "Why and when should government laboratories be involved in technology development activities?" Though the answer to this question is a long and complex one, we believe that government technology development is likely to be appropriate:

- when there are conflicting national objectives and also a need to consolidate those objectives in a given technology;
- when the regulatory environment is such that market signals become complicated by unpredictable regulations;
- when there is an interest in stimulating a technology for which there is no significant market as yet;
- when the user of the product is a government agency *and* there is a need for particularly strong ties between the technology developer and the user;
- when the scale of development and its cost is so great that not even the largest firms can expend sufficient resources.

Industrial Service Laboratories

This type of lab exists to serve the technical needs of particular industries that either perform little research or underinvest in research. For example, the construction industry performs relatively little research and, by most measures, is the least innovative among the four-digit SIC code industries. This is due largely to the structure of the industry.

Industrial service laboratories are particularly prevalent in the Department of Agriculture but are also found in the Departments of Energy, Interior, Commerce, and Defense. Laboratories of this type serve one of two basic functions. First, many conduct research with the express purpose of serving an industrial client. Second, many provide facilities and equipment for the use of industry. The first subvariety we term *industrial support centers,* the second *industrial equipment resource centers.* Examples of the former include the Spokane Research Center of the Department of the Interior and the Southern Regional Research Center of the USDA. In both cases, an industry is served that underinvests in research. The industrial equipment resource center is a type that has been in evidence for decades. One early instance (1930s) was the establishment of wind tunnels designed to contribute to the growth of the then fledgling aircraft industry. Many of the industrial resource equipment centers of today are much more sophisticated and support more varied industrial customers. The Federal Railroad Administration's Transportation Test Center is an example that includes applications of interest to a variety of industries rather than just the railroad industry. Some of the more elaborate equipment resource centers exist in laboratories that are not actually industrial service laboratories.

Viewing Policy Problems Through the New Typology

Thinking about federal laboratories in terms of the types outlined here allows for the problems and needs to be considered in terms of the niche filled by particular labs' technical foci. An important beginning point is assessing whether each of these types has a sufficient rationale. In our view, each of these roles is valid for federal laboratories, at least under specified conditions (of the sort implied in the examples provided above). The question, then, is one of balance. This leads to questions such as: "Are there too many technology development labs?" "Are there technology development labs that are poorly rationalized?" "Can technology development labs be dismantled, redesigned, or redeployed once the technology development goal is either reached or for some reason becomes irrelevant?"

Against the backdrop of the above federal laboratory organizational analysis, we can consider some of the major changes affecting the system. These include:

- movement toward commercialization and technology transfer,
- downsizing or at least transformation of defense R&D missions,
- increased interorganizational linkages among the federal labs,
- mission accretion and imperialism among labs and also competition with industry,
- recognition of new missions such as environmental cleanup,
- Increased political visibility and vulnerability.

	Pure Basic	Directed Basic	IRARL	Infratech	Tech Develop	Ind Service
Example	Kitt Observatory	PPPL	CERL	Croton USCG	USDA, Fort Collings	BOM - Spokane
Output	Sci. Knowledge	Sci. Knowledge	Mixed	Technology	Infratech	Equip Share
Function	Basic	Basic	Applied	Tech Dev.	Tech Support	Ind Support
Culture	Scientific	Scientific	Mixed	Mixed	Mixed	Market
Stability	High	High	Low	Medium	Medium	High
Frequency	Low	Low	Medium	High	Low	Medium

Table 5.1 **Attributes of Federal Lab Types**

Not all examples are extant.

MOVEMENT TOWARD COMMERCIALIZATION

If there is any single recommendation we provide here, it is that the roles of the federal laboratories within the federal system and, more broadly, the NIS should be considered in R&D policy making. This point is perhaps most dramatic in connection with the increasing commercial roles of federal laboratories. Most cooperative technology paradigm legislation makes only the broadest distinctions among labs, usually those related to size (Bozeman, 1994a). However, despite the expansion and diversification of individual federal labs' missions, it remains the case that labs have core competencies and, by implication, a lack of competency in certain aspects of technical enterprise. Considering NCRDP findings about commercialization *and* the R&D laboratory typology presented above, we recommend:

- With the exception of the very largest laboratories, which have the capacity to perform diverse technical missions capably (Bozeman, Papadakis, and Coker, 1995b), the pure basic research and the directed basic research laboratories should have minimal involvement in the growing trend toward commercial activities of federal laboratories. It is not that they are incapable of commercial work, but that the basic research missions, many of which are unique, must be preserved. Similarly, the culture that promotes basic research, a scientific culture with norms of universalism, should not be altered (as it inevitably is with the imposition of commercial demands).

- The industrial service laboratories and the technology development laboratories, although diverse in focus, should be strongly directed toward commercially relevant activities (many already have been). Those that cannot justify their existence in terms of service to a parent agency's technology needs should be evaluated by industry in terms of the utility to industry. After a period of five years or so (required to adjust to expectations) those that are not highly *rated by industry* or their federal client, such as some of the defense labs, should be decommissioned, redeployed, or radically downsized. Their resources should be transferred to those that are highly rated.

- In determining the future of technology development laboratories, criteria must be established as to the roles of these laboratories vis-à-vis industry laboratories. They should not duplicate or drive out industry technology development, rather they should focus their technology development activities on the following:

1. projects of a scale and magnitude beyond the means of individual private firms, even the largest;
2. technologies for which the market is a long-range one and in the national interest;

3. technologies that are vital to the national interest but for which the sales possibilities are quite limited (e.g., in the past, nuclear weapons technology);

4. technology development for industries that have limited R&D capabilities (e.g., construction) but that represent a significant public interest value or economic value. In these cases, efforts should be made to determine the reasons why the industry has limited capacity and then to develop that capacity, perhaps through coproduction of technology;

5. in those cases where none of the above conditions pertain and the technology is not being developed for specific use by a mission agency, the technology development lab or the particular program should be reevaluated and either closed down or redeployed.

- If there is any lab type that is a candidate for expanding its numbers, it is the infratechnology lab. These labs can be vital in industries that are in need of support structures of the sort that are not feasible or that cannot be economically provided for by any single performer or group of performers within the industry. A prospective study should be commissioned to determine on an industry-by-industry basis the desirability of launching new infratechnology laboratories. Any new laboratories of this sort might well be financed by redirecting money from technology development laboratories or IRARL labs that are outmoded. Infratechnology labs should not be established unless there is clear industrial support, perhaps tangible support through cost-sharing.

- Whereas the IRARL labs are of most value in their "bridging" activities, a significant proportion of their work should be designated for cooperative research and coproduction.

- IRARL laboratories should be provided (or their mission agencies should be provided) with discretionary funds that can be used, independent of sector or ownership, to further the development of technology to the point of market impact (see Bozeman and Pandey, 1994, for discussion of a similar role for the Japan Research and Development Corporation).

Downsizing of Defense

There is at least some impact from this trend throughout the federal laboratory community, with the greatest impact being in those technology development laboratories that have been created to provide weapons technology. With the possible exception of the DOD laboratories, those laboratories that are the most "defense-dependent" should be reassessed to determine their possible utility as something other than a development operation for defense technology.

Various NCRDP findings suggest that the possibilities for shifting defense-related activities depends in large measure on the extent to which laboratories already have historic missions that extend beyond defense as well as upon the level of capacity and slack resources of the labs (Bozeman and Crow, 1990, 1995). In particular, the transition from technical activity with a classically sheltered or guaranteed market to technical activity aimed at producing technology or knowledge to be appropriated in private markets is one that is likely to fail. The effort to take scientists and engineers who have for years produced the science and technology behind weapons systems and put them suddenly into the service of industry, producing technical output to enhance competitiveness, is largely misguided. With regard to those technology development and intermediate-range applied research labs that are the most defense-dependent, we recommend:

- Assess likely defense technology needs given alternative plausible scenarios, identify unique, largely irreplaceable technical capabilities, and close down programs that do not fit within those categories. This will lead to a radical downsizing beyond what is commonly being proposed.

- Do not try to reassign weapons technology development personnel to civilian technology objectives, unless there is a substantial retraining effort and unless these personnel are assigned to new or preexisting laboratories that have a commercial, civilian technology mission.

- For those personnel not easily reassigned (because of lack of interest or lack of demand) provide a generous early retirement or severance option (which will be recompensed in the near-term by a "defense technology development dividend." One option is to keep budget outlays stable for one to two years but to devote the expenditures entirely to severance. In the short term, this is a good investment, and it is better than seeking to redirect outmoded defense technology development operations into new civilian enterprises.

INTERORGANIZATIONAL LINKAGE AMONG THE FEDERAL LABS

During the past two decades there has been a considerable increase in interorganizational linkages among the federal laboratories (Bozeman and Crow, 1988). While "work for others" and interlaboratory collaboration has for years been commonplace for many of the DOE multiprogram laboratories, other federal labs have begun to move beyond a traditional localism and isolation. In most respects, this is a positive trend, and we recommend that it be accentuated.

- Department or agency affiliation should not be a barrier to laboratory teaming. Increasingly, federal laboratories should be viewed as national resources, and the more expansive technical missions of federal labs should deploy resources from multiple laboratories with diverse affiliations. This will require a redesign of budgeting for R&D which, historically, has been conducted through the lab parent and according to the lab parent's mission. The multiprogram "national labs" already have a perspective that views federal agencies as customers. Increasingly, mission-oriented federal agencies must be permitted to draw from a wide range of government and non-government laboratories rather than just their traditional suppliers.

- Just as there are "virtual corporations" springing up, the federal laboratory systems should be viewed as a source of "virtual laboratories," drawing capabilities from diverse sources. In order to do so, however, a much better inventory of lab skills and capacity will be required.

Mission Accretion and Imperialism: Competition with Industry

The increasing technical cosmopolitanism of federal laboratories, while in many ways quite healthy, has in most respects also had some unfortunate consequences. As competition for resources increases or is increasingly perceived, the "natural" reaction of organizations is mission accretion or imperialism. This takes two predominant forms: competition with other federal laboratories and competition with industry. Neither is inherently bad so long as it is rationalized. The problem is that there is little rationalization to the competition; instead, it is driven by the resourcefulness and aggressiveness of particular individuals and leaders of particular labs. One result is that signals are becoming mixed as to which laboratories have what technical responsibilities. If "everyone" is in the clean vehicle business, the superconductivity business, or the MRI business, then resources are unlikely to be used most efficiently, and potential industrial partners (or other users of labs' technical output) are less likely to know where to go for particular types of information and technology. Similarly, it becomes less clear when competition with industry is potentially useful and when it is harmful. Thus, we recommend:

- Federal laboratories need to be reconfigured such that their technical capabilities and national needs determine their activities to a greater extent and the missions of parent agencies to a lesser extent. This is especially important for pure basic research and applied basic research labs, less so for technology development labs.

- "Competition" with industry should be encouraged only when the criteria articulated above are met, not just because the lab is seeking new missions to enhance its survival.

RECOGNITION OF NEW MISSIONS: ENVIRONMENTAL CLEANUP

A different issue from mission imperialism is the rise of entirely new missions. During the past decade, one of the signal events in federal laboratories has been the increasing significance of public domain missions not related to defense. Thus, for example, environmental research and hazardous waste mitigation are clearly "public goods" in their economic character in much the same way that defense has been. That is, the benefits are not clearly divisible, and there is no efficient market for many of the technical outputs related to the environment. We recommend:

- Particular attention needs to be given to those missions that are neither defense-related nor of short-term market relevance, the "other" public goods and public interest laboratory missions. Rather than viewing defense conversion and other laboratory realignment efforts entirely in commercial terms, it is important to give emphasis to emerging public interest missions. One reason for this is that federal laboratories are in a unique position to fulfill such missions of "high publicness" and often do so with a high degree of effectiveness (Crow and Bozeman, 1987b).

INCREASED POLITICAL VISIBILITY AND VULNERABILITY

A clear trend affecting federal laboratories is an increased political visibility and at least a perception of vulnerability. This is not a healthy atmosphere in the long term. While it is certainly understandable that the role and scope of federal laboratories is now being reexamined, it is important to make sure that this process proceeds expeditiously. Virtually all the NCRDP work pertaining to the effects of political change has shown that labs with a high degree of external political instability do not perform as well and often take steps that are in their own perceived self-interest but against the collective interest; mission imperialism is a good example (Bozeman and Loveless, 1987; Bozeman and Crow, 1990). While we have no NCRDP-based recommendations pertaining to political visibility and vulnerability, we do suggest that any contemplated realignment and redirection be undertaken as swiftly and unequivocally as possible yet also deliberately and prudently. In our visits to government laboratories, the sense of dangling in the wind has been palpable.

During the past decade, federal laboratories have been under the microscope. The changes in the structure and composition of the federal laboratory system have been considerable. With few exceptions, however, the changes have been ad hoc, uncoordinated, responsive to temporal events rather than systematic changes following the course charted by any of the several study panels examining federal labs. The question, then, is whether change could be more systematic and sensitive to strategic needs. Several strategic issues are considered in the concluding chapter. The recommendations we provided in this chapter are summarized, along with those provided throughout the book, in a table at the end of the final chapter.

Before turning our attention to additional recommendations, we examine in chapter 6 empirical findings from NCRDP studies of the performance of federal laboratories, especially with respect to technology transfer and commercialization activities.

CHAPTER SIX

Federal Laboratories and Their Performance in the National Innovation System

In the wake of the perceived crisis in U.S. competitiveness (MIT Commission on Industrial Productivity, 1989; National Academy of Sciences, 1978; National Governors' Association, 1987; President's Commission on Industrial Competitiveness, 1985), many core assumptions about science and technology policy began to be reconsidered in the late 1980s and early 1990s, including the government's role in spurring industrial innovation, particularly through its investment of more than $25 billion annually in its formal R&D laboratory organizations. As suggested in chapter 2, the assumptions of the market failure and mission paradigms, the dual columns upon which modern U.S. science and technology had been built, began to be challenged in the 1980s by a new approach emphasizing federal government and industry R&D partnerships.

A significant number of cooperative paradigm policies are now in place, although weakened since 1994 by Congressional Republicans' concerns that cooperative policies are just industrial policy in disguise. But the late 1980s and early 1990s ushered in several cooperative development policies including changing patent law to expand the use of government technology (Patent and Trademark Laws Amendment, 1980), relaxing antitrust, promoting cooperative R&D (Link and Bauer, 1989; National Cooperative Research Act of 1984), establishing research consortia and multisector centers (Smilor and Gibson, 1991), and altering the guidelines for disposition of government-owned intellectual property (Bagur and Guissinger, 1987; Gillespie, 1988; Bozeman, 1994a; Coker, 1994; De la Barre, 1985; U.S. General Accounting Office, 1989).

The cooperative paradigm policies that have attracted the most attention are those pertaining to domestic technology transfer, especially the use of federal laboratories as a partner in the commercialization of technology (Herrmann, 1983). Previously aloof from commercial concerns and indeed prohibited by law from developing technology specifically for private vendors, the federal labs' mission, tenor, and climate has to some extent been changed by the legislation of the 1980s. The intellectual property dictum "if it belongs to everyone, it belongs to no one" began to take hold as the government labs increasingly moved from a focus on public domain research to a mandated role as a technology development partner to industry.

The cooperative paradigm's zenith may have been reached between 1993 and 1995. In 1992 the Clinton-Gore campaign's (Clinton and Gore, 1992) science and technology policy manifesto brought to the policy debate a strong set of cooperative paradigm prescriptions. However, only a handful of the policies urged in the campaign document were put forth, and even fewer where ultimately implemented. Before science and technology policy had a chance to move to the "front burner" of the policy agenda, the 1994 elections brought a new Republican congressional majority dead set against cooperative paradigm policies and often suspicious of traditional mission paradigm policies. The Republican Congress wasted no time translating its overall zeal for putting public programs on the chopping block into specific cutback proposals. Almost all of the nation's science and technology policy outlays are part of the "discretionary budget," that part of the budget not based on entitlements, not related to interest on the general debt, and not having "automatic" funding formulae. One expression of congressional Republicans' commitment to less government was strong support for dismantling the Departments of Energy and Commerce, two of the chief apostles of the cooperative paradigm and the mission paradigm in science and technology policy. While congressional Democrats, the administration, and a few sympathetic (and often politically self-interested) Republicans were successful at beating back the elimination of these departments in 1995 and 1996, many federal laboratories remain on notice at the close of the 1990s.

Between 1995 and 1997, a siege mentality pervaded many federal laboratories (possibly excepting some Defense Department laboratories). Many labs sought feverishly to prove, via "metrics," expert testimony, or anecdotes that their contributions to the nation's economy and technological advance were sufficient to warrant continued funding for the labs. Some federal labs, reading the political tea leaves a bit differently, began reducing their industrial partnerships and deemphasizing the lab's commercial role, focusing instead on public domain science. But neither strategy seemed adequate to stem cutbacks as thousands of federal laboratory personnel were given early retirement packages or laid off.

In the wake of the political controversies at the center of which the federal laboratories are, it is appropriate to consider in some detail the contribution of the laboratories to industry and to examine industries' objectives and strategies for working with federal laboratories. In today's political climate there are surely many ideologues who simply do not care whether federal laboratory and industry partnerships have proved beneficial; if one views federal laboratories as having no role in promoting industrial competitiveness, then evidence that the role is played well falls on deaf ears. But even today most politicians are pragmatic and evidence of effectiveness is a useful, even if not predominant, input into the policy-making equation.

Technology Transfer Performance at Federal Laboratories

This chapter provides evidence about federal laboratories' commercial technology transfer activities. We began with NCRDP studies (Bozeman, 1994a; Bozeman and Coker, 1992; Bozeman and Fellows, 1998; Coursey and Bozeman, 1992) of federal laboratories, examining their motives, strategies, and performance in technology transfer. But data from the federal laboratories tell only part of the study. Examining more recent data and studies from the NCRDP (Bozeman, Papadakis, and Coker, 1995; Bozeman and Papadakis, 1995; Bozeman and Rogers, in press; Bozeman, 1996b), we consider results reported by industrial partners to federal laboratory projects. After examining this evidence it should be a bit clearer whether the cooperative paradigm optimists, such as Alan Schriesheim and others (Schriesheim, 1990; Kearns, 1990; Krieger, 1987; Chapman, 1986; Conference Board, 1987), have a compelling case.

Which Federal Laboratories Are Involved in Technology Transfer?

As a result of poor federal R&D policy and policies based on noninstitutional models, technology transfer is now a mandated activity for many federal laboratories. Examining results from NCRDP studies (Bozeman, 1994a; Bozeman and Coker, 1992; Crow and Bozeman, 1991), we can gather some ideas abut the level of participation in technology transfer. This is particularly important because early evidence (U.S. General Accounting Office, 1989) suggested that technology transfer legislation had not been fully implemented in federal laboratories, that some labs were not engaged in technology transfer at all, and that others had not devoted the resources to technology transfer required under legislation. Historical data from the 1991 NCRDP research indicated that about one quarter of federal laboratories examined ($n = 212$), from a representative sample of all federal R&D labs in the U.S., were not at that time involved in technology transfer. While this figure may have decreased since 1991, many small laboratories are not required to engage in technology transfer activities, and despite the apparent level of effort in technology transfer, the percentage of noninvolved laboratories may not have changed much. It is also worth noting that in 1991 very few of the laboratories reported that technology transfer was a major mission. Only two of the labs (both from the Department of Agriculture) reported that technology transfer was *the* major mission of the laboratory (Bozeman, 1994a). About half the laboratory directors reported that technology transfer was "an important mission" (Bozeman and Crow, 1994a).

Lab Window 6.1 "All Roads Lead to Rome"

Rome Laboratory

http://www.rl.af.mil/Technology/rl%2dtechno%2dmain.html

Rome Laboratory emphasizes technology transfer—the sharing or transferring of information, data, hardware, personnel, services, facilities, or other scientific resources for the benefit of the private or public sector. At Rome Laboratory our research and development in the areas of signal, speech, and image processing, communications, electromagnetics, photonics, computational sciences, reliability, maintainability, and testability will lead to many future products that will greatly benefit both the military and the commercial sectors.

Rome Laboratory emphasizes technology transfer—the sharing or transferring of information, data, hardware, personnel, services, facilities, or other scientific resources for the benefit of the private or public sector.

Congress passed specific legislation encouraging technology transfer in 1980. Since, then, Congress has continued to strengthen technology transfer through additional pieces of legislation. The White House has also initiated new programs emphasizing the transfer of technology to stimulate the economy and to improve the international competitiveness of American industry.

Rome Laboratory's mission of research and development in the information technologies for the Air Force and DOD is ideally positioned to also satisfy congressional technology transfer legislation. Rome Laboratory's Information Technology developments are by their very nature applicable to both the military and the commercial world—they are "dual use."

Since its establishment in 1951, Rome Laboratory has maintained an excellent track record of developing dual-use technologies and making them available to the private sector. Rome Laboratory has touched almost everyone in one way or another—examples include learning aids for children, compact discs, computer memory, security systems, fiber optics, and satellite communications, to name a few.

Rome Laboratory is continuing its outstanding multifaceted technology transfer program. All types of technology transfer agreements authorized by Congress are used by Rome Laboratory. The principal types of agreements used are (1) Cooperative Research and Development Agreements (CRDAs) with industry, academic, and state and local government agencies; (2) Education Partnerships with academic institutions; and (3) Patent License Agreements with private industry. Other activities include the issuance of grants and cooperative agreements, formation of consortia and regional alliances, and answering requests for technical assistance.

Another way of determining the importance of laboratories' technology transfer activity is in terms of the budget devoted to technology transfer. Federal laboratories participating in Phase III—NCRDP research ($n = 350$) spent an average of about \$191,000 annually (1992 dollars) on technology transfer activities, some 6.28% of their total R&D budgets (Bozeman, 1994b). This percentage, however, distorts the wide range of variance in budget allocations to technology transfer (standard deviation = 11.48), with one lab reporting virtually its entire budget devoted to technology transfer, and many reporting none of their budget devoted to that activity.

On balance, the evidence suggests that technology transfer and commercial activity remain a relatively modest part of most federal laboratories' missions. While it is difficult to provide an exact accounting (since budgets and personnel sometimes do not have technology transfer line items), it seems likely that few labs devote as much as 10% of their resources to technology transfer and ancillary activities such as technical assistance to industry.

TECHNOLOGY TRANSFER EFFECTIVENESS: CONCEPTUAL AND MEASUREMENT ISSUES

The key issue, of course, it not the *intensity* of laboratories' involvement in technology transfer, which varies by lab mission and lab type, but the *effectiveness* of that activity. The problem lies in the fact that it is usually much easier to assess the incidence of activity than its impact. This has led to a tendency to count Cooperative Research and Development Agreements (CRADAs), to parade licensing agreements, and to compile exhaustive lists of industrial partners. But none of this tells us whether the technology transfer hubbub is more than show.

It is not easy to measure the impact of technology transfer. It is not even easy to conceptualize its effectiveness. An early NCRDP effort (Bozeman and Fellows, 1988) tried to develop conceptual models of technology transfer inductively from a series of case studies. The dearth of models of technology transfer effectiveness makes that model worth reviewing here, especially since it influenced empirical evaluations (Coursey and Bozeman, 1992; Bozeman and Pandey, 1994; Bozeman and Coker, 1992; Bozeman and Crow, 1991a).

The Bozeman-Fellows technology transfer effectiveness model was developed from a series of case studies performed at Brookhaven National Laboratory (BNL) (Bozeman and Fellows, 1988). The initial aim of the case studies was to document the range of commercially relevant technology transfer and assistance work provided at large multiprogram laboratories. Each of the case studies exemplified a major form of laboratory technology transfer activity including transfer of physical technology or prototypes ("Superconducting Magnets Case"); transfer of technological processes ("Polymer Insulation Case"); transfer of "know-how" or craft ("Remotely Operated Robotic

Arm Case"); resource and equipment sharing ("National Synchrotron Light Source Case"); and provision of technical assistance ("Three Mile Island Corrosion Research Case"). As a result of the case study research, it became apparent that no single, unified approach to technology transfer effectiveness is adequate—the activities and goals are simply too diverse. To address this diversity and to help shape technology transfer assessment efforts, various "models" for understanding and measuring technology transfer success were developed through the BNL case.

"Out-the-Door" Model of Technology Transfer Effectiveness

Until recently, the model embraced by federal laboratories and others wishing to *measure* technology transfer success was the "out-the-door" model. The assumption of the out-the-door model, depicted in figure 6.1, is that transfer itself equates with success. Naturally, most parties involved in technology transfer realize that "dud" technologies have the potential to actually harm adopters and that even "sweet" technologies may have no significant impact if they are adopted but then poorly implemented. But from a practical standpoint the out-the-door approach is useful. Not only is it quite simple, it is usually easy to determine whether a technology has been adopted, and it has the important advantage of basing the evaluation criteria on factors largely under the lab's control. Some would say that the lab is at least partly culpable if it transfers technologies to companies who have inadequate capital, manufacturing ability, or market savvy to make a good technology into a good, profitable product. But that is a high standard and requires market forecasting expertise, which is in short supply at the federal laboratories (and not widely available in industry). The implicit argument of the out-the-door model is that it is the laboratory's job to create technologies or applied research attractive to industry, but it is industry's job to make them work in the marketplace.

One of the Brookhaven case studies illustrates the dilemma of the out-the-door model. The remotely operated robotic arm was a significant technological innovation; it was a technology that worked flawlessly, and it was transferred to a newly formed corporation. The remotely operated arm is an enhancement to robot technology that enables the operator to "feel" what the robot is touching and to operate the arm of the robot for distances of up to one mile—an obvious benefit when handling hazardous materials. The technology was transferred to a former Brookhaven scientist who had worked on the project and thus had ample technical expertise. But the technology was not a "winner" in the marketplace. Due to the scientist-entrepreneur's limited business experience and modest capital, the company's resources were inadequate for successful commercialization of the product. But from an out-the-door perspective, the laboratory was effective.

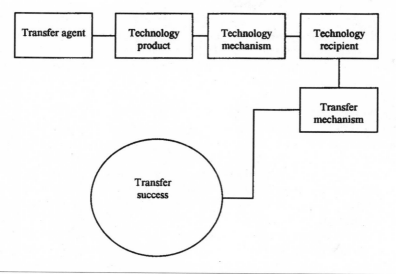

FIGURE 6.1   "Out-of-the-Door" Definition of Effectiveness

Market Impact Model of Technology Transfer Effectiveness

The *market impact* model, as the name implies, assesses effectiveness according to the commercial success of the transferred technology or information. The advantage is that it is a richer notion of success—there is little appeal to technology transfer that proves commercially and instrumentally barren—but it entails conceptual difficulties. If the transfer is not commercially successful, is it because the product or process transferred is of limited value, because the transferring agent has not taken the actions necessary to ensure its success, or because of the recipient organization's problems in development, manufacture, marketing, or strategy? These problems are particularly troublesome if one plans to use an effectiveness evaluation to recommend action and change.

Figure 6.2 shows that the market impact model is quite similar to the out-the-door model, except for the critical assumption that transfer success requires that the transferred product or process contribute to the firm's profitability. Most agree that market impact is the acid test of technology transfer. Still, there is the problem that the laboratory does not control many vital aspects of market success. Thus, in the Superconducting Transmission Line Case at Brookhaven, the value of the technology was greatly diminished by rapid changes in energy costs—a sad story that could be told by many energy technology inventors and entrepreneurs.

Clearly, most of the objectives and nearly all the rhetoric of cooperative paradigm technology policy centers on a market impact concept of success. The cooperative paradigm is rationalized in terms of its potential to contribute to competitiveness, and thus technology transfer with little market result has no place in the paradigm.

The Political Model of Technology Transfer Effectiveness

During various on-site interviews conducted by NCRDP researchers, laboratory officials have on many occasions made direct or, more frequently, indirect reference to the political payoffs expected from technology transfer activities. That is, technology transfer is viewed as a way to enhance political support rather than as a means of creating direct resources or contributing to industrial competitiveness. Thus, it is an instrumental value, a means to an end. Figure 6.3 depicts the *political model.*

There are at least three possible avenues to political reward, each depicted in the model. In the least likely of scenarios, the lab is rewarded because the technology it has transferred has considerable national or regional socioeconomic impact and the lab's role in developing and transferring the technology is recognized by policy superiors, and in turn the lab is rewarded with increased funding or other resources. In this view, the political and the market impact models are highly complementary. This scenario is not unprecedented but does not commonly occur. In the first place, few technologies have such an impact. But even when there are huge impacts from technology transfer, often the laboratory role is not evident to policy makers, or the

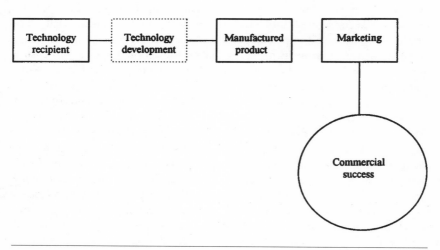

FIGURE 6.2   Market Definition of Effectiveness

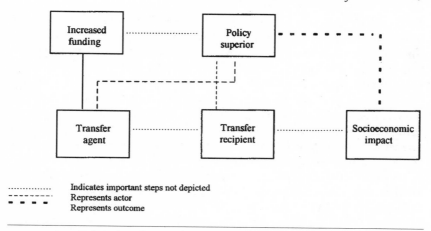

FIGURE 6.3   Political Definition of Effectiveness

policy makers simply may not provide the expected "reward." Budgeting processes usually do not work in a way to reinforce such expectations.

Another way in which the political model may yield resource results for the laboratory is through the recipient firm, the firm benefiting from the technology transfer, communicating to policy makers its pleasure with the lab interaction. The policy maker then, in turn, rewards the lab for being a "good industrial partner." This is a bit more plausible than the first scenario in that it matches the reality or constituency-based politics and the practicalities of "pork barrel" funding of science and technology resources (Bozeman and Crow, 1990).

Probably the most common and realistic rationale under the political model is for the lab to be rewarded for the *appearance* of active and aggressive pursuit of technology transfer and commercial success. While the direct returns are few and the possibly many indirect returns difficult to measure, much bureaucratic behavior seems to support this view. Often federal laboratories are as active in publicizing their technology transfer and economic development activities as in actually doing the transfer work. Furthermore, the obsession with CRADA-counting and other input measures could be viewed as rooted in the political model, based on the assumption that "good numbers" will lead to political reward.

The Opportunity Cost Model of Technology Transfer Effectiveness

One abiding truth about federal laboratories' technology transfer activity is that it is only one of many technical activities occurring and usually not—at least in the lab scientists' and technicians' view—the most important. Further, many other activities of federal laboratories are vitally important.

Transferring technology takes its place alongside contributing to the advance of basic research and scientific theory, providing equipment and infrastructure for the growth of scientific knowledge, training scientists and engineers, and ensuring the nation can perform its defense, national security, public health, and energy missions.

Consistent with the institutional design principle, the opportunity cost model recognizes that resources going into technology transfer might, at least theoretically, have been used for other purposes. Thus, the more stringent evaluative criterion under this model compares the value of technology transfer to other possible uses of the resources. Related to this is that the opportunity cost model considers that technical activities in laboratories affect one another and, thus, examines the impact of technology transfer on the rest of the lab's technical portfolio. Figure 6.4 depicts the opportunity cost model.

The model views laboratory goals in terms of three activities, technology transfer, knowledge transfer (i.e., generation and dissemination of fundamental scientific knowledge), and "other," which might include administrative support, capacity-building, and, generally, scientific and technical infrastructures. The opportunity cost model is the most difficult one to implement. Some of the problems it poses: (a) determining the net benefit of basic research (a classic problem still unresolved by policy analysts), (b) analytically separating the obviously intertwined activities of science, scientific support, and technological delivery, (c) determining a wide variety of potential interaction effects. Nevertheless, the opportunity cost model is at

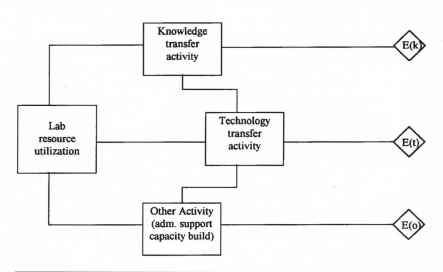

FIGURE 6.4   "Opportunity Cost" Definition of Effectiveness

least important as a conceptual tool inasmuch as it encourages analysts to take a more global view of the laboratories' activities.

Technology Transfer Effectiveness: Laboratory Data

The findings we report concerning technology transfer effectiveness are from two different phases of the NCRDP and from two different perspectives. One phase focused on data derived from the laboratories, the other on data derived from the laboratories' industrial partners. In this section we review findings from the study of federal laboratories; in the next we examine findings obtained from companies who have worked with the laboratories.

In both phases, the implicit effectiveness models examined were the market impact and out-the-door models. However, the laboratory-based data is more oriented to the out-the-door model, whereas the industry-based data is more oriented to the market impact model.

The laboratory-based data includes the number of licenses and patents (for 1990) as well as several self-rating indicators. Licensing patterns among the federal laboratories indicate an average of 1.5 licenses. However, the average is perhaps deceiving. A few laboratories hold an extremely large number of licenses and 77 of the federal laboratories have none at all. By this indicator, a few laboratories have "cornered the market." Similar to licensing, the average number of patents reported by the laboratories is not vast, only 4.5, but, again, there is a highly skewed distribution, reflected in the fact that the median number of patents for all laboratories is just one. While 50 (35.2%) of the laboratories reported no patents, 8 (5.6%) reported more than twenty.

Laboratory directors were questioned about the benefits the lab had received from technology transfer activity. These questions included the level of profit for both the lab and individual lab scientists. The level of perceived benefit from these two indicators of profit—for the lab and for the individual—indicate that the benefits are modest. With the *average* score indicating that the benefits are insignificant, we see that only a few labs have experienced much in the way of remuneration, either for the lab or for the scientists and engineers involved in technology transfer. An examination of the original data indicated that only 2 (1.3%) of the laboratories considered profit for the lab as the most important benefit, and only 11 (7.5%) saw it as a major benefit. Similarly, only 5 (3.4%) considered profit for the lab scientists and engineers as the major benefit and only 14 (9.5%) as a major benefit.

How Does Motive Relate to Technology Transfer Effectiveness?

It seems reasonable that the degree and type of success in technology transfer might relate to the motive for pursuing the mission. Laboratory directors were asked to evaluate the importance of the following motives for their laboratory's technology transfer activity:

- Legislative requirements or statutory mandate
- Economic development emphasis of the lab
- Outgrowth of cooperative R&D or research consortium
- Participation in industry-university or government-university research center
- Exchange of technical information or personnel
- Hope to increase lab's or parent agency's budget
- Scientists' and engineers' personal satisfaction upon seeing their ideas or technologies developed
- Scientists' and engineers' interests in entrepreneurship and personal wealth

One might assume that emphasis on the legislative mandate as a motive for technology transfer would be negatively related to success, simply because such a desultory motive ("we're in the technology transfer business because we have no choice") would rarely lead to success. However, this study shows that there is no significant relationship between following a mandate and the transfer effectiveness measures.

Three motives seem, according to correlational analysis, especially important (Bozeman and Coker, 1992). The economic development motivation is significantly and positively associated with *all* the success measures. The entrepreneurial motivation is significantly and positively associated with both the benefit to the lab and to the individual scientists along with the number of licenses. Perhaps the most interesting finding, at least from a public policy standpoint, is the positive relation between technology transfer success of various types and the motive of participating in a research center. That is, labs that view technology transfer as an outgrowth of center participation are more likely to have success. This seems to imply that the attention paid to government-industry research centers in past years is likely to provide substantial technology transfer benefits.

Do Laboratories' Other Missions Complement Technology Transfer Effectiveness?

Few laboratories exist for the sole purpose of technology transfer. Does the laboratory reporting basic research as a major mission enjoy greater or lesser technology transfer success? Are technology development laboratories more successful? Are laboratories with a greater number of missions more successful?

The mission variables are based on questionnaire items asking the laboratory directors to identify the significance of each of the following missions:

- Basic research (knowledge for its own sake without any particular application in mind)
- Precommercial applied research (focused on bringing new products and processes into being but not directed toward a specific design)
- Commercial applied research (focused on product or process with specific design in mind)
- Development (developing existing prototypes, modifying existing products/processes or applications engineering)
- Providing technical assistance to government agencies
- Providing technical assistance to private firms and industrial organizations
- Providing technical assistance to this laboratory's parent organization

One might expect some lab missions to be compatible with technology transfer and others not. For example, one might assume that basic research, science for its own sake, might be at odds with science and technology for commerce's sake. NCRDP findings give some modest support to this expectation in the relationship between basic research and out-the-door success (Bozeman and Coker, 1992; Bozeman, 1994a). Similarly, one might assume that laboratories involved in applied research and technology development would likely be more compatible with the technology transfer mission. But this assumption needs to be refined. There is no significant relation between a technology development mission and technology transfer success. However, several significant relationships exist between the two applied research missions, commercial applied and precommercial applied. Technical assistance missions of laboratories relate to technology transfer effectiveness, but with opposite effect according to the object of the assistance (Bozeman, 1994a). Technical assistance to a parent agency is negatively related to technology transfer effectiveness in terms of licenses and profit for lab scientists and engineers. Technical assistance to industry is positively related, at least for out-the-door success and market impact.

WHICH TECHNOLOGY TRANSFER STRATEGIES ARE THE MOST EFFECTIVE?

Presumably, some technology transfer strategies work better than others. The respondents evaluated the degree of success of each of the strategies listed below:

- On-site seminars and conferences
- Fliers, newsletters, or other mailed correspondence

- Person-to-person contacts of our scientific and technical personnel with persons in technology recipient organizations
- Presentations at scientific meetings sponsored by professional organizations
- Presentations at scientific meetings sponsored by government organizations
- Membership in research consortia, university or government centers
- A central office with responsibility for technology transfer
- Encouraging informal, on-site visits
- Personnel exchanges
- Cooperative R&D (as a technology transfer strategy rather than for other possible purposes)
- Contractual relations for direct R&D funding between our lab and the organization receiving the technology
- Permitting persons from other organizations access to our laboratory's equipment and facilities.
- Sales or gift of patents, copyrights, or licenses

Membership in research centers seems to have considerable payoff in terms of technology transfer success. Four of the five effectiveness indicators are significantly and positively associated with this strategy (Bozeman, 1994a). Similarly, cooperative R&D as a strategy also seems to have positive outcomes for technology transfer success. This is most often seen in regard to getting the technology out the door and reaping some market success.

The sale of patents and licenses is positively associated with each of the effectiveness indicators. This is perhaps not surprising since this particular strategy may be a good reflection of how serious the laboratory takes its commercial technology transfer activities.

The use of a central office for technology transfer doesn't seem to make much of a difference, but the single significant relationship is worth noting. The strategy of having a central office for technology transfer seems to result in pecuniary benefits to the lab's scientists and engineers. It may, perhaps, provide the administrative support needed for these part-time entrepreneurs to realize economic benefit from their work.

Finally, out-the-door-success is the type of technology transfer most "strategy-sensitive." Nine of the various strategies are significantly and positively related to out-the-door success, possibly because it is the least stringent success criterion. By contrast, the number of licenses issued seems to be largely independent of strategy. This further underscores the importance of the findings about the relationship between the research centers' strategy and licensing success. It is one of the few strategies that seems to work.

Industry Partners' Perspective on Federal Laboratories

As we mentioned in chapters 1 and 3, various policy developments from the cooperative technology paradigm have altered the opportunities for industry use of federal technology and collaboration with federal labs. But to what extent does industry benefit from the intended transformation of federal labs into science and technology shopping malls? To address this question, we focused Phase V of the NCRDP (Bozeman, Papadakis, and Coker, 1995b; Bozeman and Papadakis, 1995; Bozeman and Rodgers, in press) on assessing the impact of federal laboratories' science and technology on their industrial partners. These industry-owned laboratories examined in Phase V were largely private technology labs by type, yet a few were hybrid technology or hybrid science and technology labs.

It is important to note that the sampling approach here was very different from that of the other NCRDP studies. In the first place, neither the lab nor the company was the sampling unit. The technology/knowledge transfer interaction between the company and the lab was the focus. Nor was there an attempt to sample randomly. We focused on laboratories prescreened as particularly active in technology transfer work. While 27 major federal laboratories, representing the public science, public science and technology, and public technology types of labs participated, more than 70% of the interactions were with the multiprogram national laboratories of the DOE.

It was not our objective to necessarily focus on DOE labs, but by most measures these labs are among the most active in technology transfer and commercialization.

Characteristics of Federal Lab-Industry Interactions

Since the sampling strategy focused on those federal laboratories particularly active in working with industry, there is no reason to believe that the characteristics of these interactions are representative of those in all federal laboratories. Nonetheless, the data seem to provide a useful snapshot of the activities of the commercially active laboratories.

The type of interactions and their relative frequency is depicted in figure 6.5. There are 330 interactions reported (since some projects involved more than one type of interaction, the totals for type of interaction add up to more than the number in projects). There are three dominant categories: CRADAs, technical assistance, and cooperative R&D other than CRADAs. A total of 56% of the 1992–1993 projects are CRADAs. While all of the projects in this study have start dates after implementation of the Cooperative Research Act of 1984 (which enabled CRADAs), only 28% of the projects started before 1990–1992 are CRADAs. This implies that the use of the CRADA as a partnership vehicle grew dramatically in the early to mid-1990s. No doubt, this is in part due to quite clear directives from the two secretaries of energy in office between 1990 and 1996.

TYPE OF R&D AND WHO PERFORMED

Companies were asked to indicate the type of R&D or technical activity performed by their laboratory partners. The results appear in figure 6.6. The results are a bit surprising. While many of the projects (120) involved directed basic research performed by federal labs, 83 involved company-performed directed basic research. Only about 11% of industrial labs are significantly involved in basic research (Bozeman and Crow, 1988). One might speculate that this finding is an artifact of the limited construct validity of "basic research," but evidence by Link (1996) indicates that industry responses tend to have stable, internally consistent meanings for the term. These meanings coincide, to a considerable degree, with others' usage. More likely, the surprising incidence of company-performed basic research is due to the unrepresentative character of the companies involved in federal laboratory partnerships. This finding is not due to the fact that these are chiefly large businesses. The median size of the companies' R&D personnel is 14% of total personnel, indicating that these are certainly research-intensive companies.

Directed basic research is the most common category of technical activity for the labs (n = 120, 58%). For companies involved in interactions, the

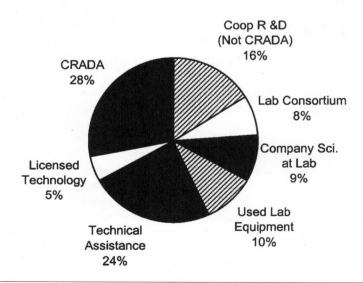

FIGURE 6.5   Types of Company-Lab Interactions

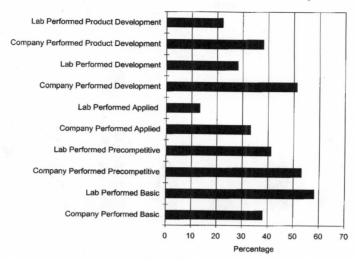

FIGURE 6.6   Tech. Activity: Company v. Lab

most common categories are precommercial R&D (*n* = 113, 53%) and development (*n* = 105, 51%). Applied research is not a major category for the federal labs and has the lowest frequency for the companies as well.

Why Do Companies Work with Labs?

Federal labs have "return business." As can be seen from figure 6.7, as many as 91 (43%) respondents indicated that they had previous projects with the specific federal lab in question. Only 59 respondents (28%) had no previous interaction with the lab. With 218 instances of such interaction prior to the forging of the current project, it appears that informal and professional contacts are important.

When asked about the reasons important "in your company's decision to work with the federal lab on this project," the most common response (127) was that the skills and knowledge of the federal lab's scientists and engineers were important. Almost as important (105) were the unique facilities of the lab. While many of the companies had previous experience with the lab (see above), prior interaction was not reported as a major reason for choosing to work with the lab on the project in question (though contacts with the labs' scientists and engineers were important [86 affirmative responses][see figure 6.8]).

There are three major objectives for companies' interaction with the federal laboratories: engaging in strategic precommercial research (111),

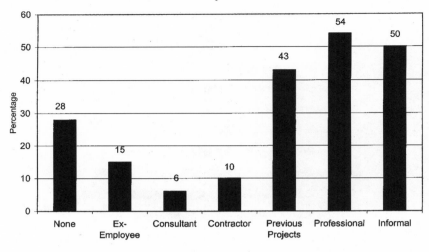

FIGURE 6.7   Previous Experience with Lab

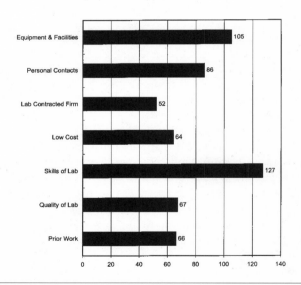

FIGURE 6.8   Reasons for Working with Lab

interest in access to unique lab expertise (98), and desire to develop new products and services (90).

There is considerable difference of opinion as to how well the laboratories can assist industry (Roessner and Bean, 1991, 1993). Some argue that the labs have more potential as partners in precommercial research than as codevelopers of new products. However, it is worth comparing the characteristics of companies motivated by precommercial R&D to those of firms motivated by the promise of new product development. Our data indicate that the objectives of precommercial R&D and new product or service development are certainly not mutually exclusive, since for 31% of the interactions both were objectives. But if we compare only those reporting precommercial R&D objectives and those reporting only new product objectives, we get a picture of firms and interactions quite a bit different from one another.

Firms collaborating with federal laboratories in order to enhance their precommercial R&D (but not for product development) are smaller firms, the dominant form of R&D labs in the U.S. NIS. The most important distinction between firms chiefly oriented (in their interaction with a federal laboratory) toward new product development and those not so oriented, is that the companies motivated by the development of new products report only about one-tenth of the benefits of all other companies. These findings are complex and, as we suggest below, require some sorting out.

Benefits and Costs of Federal Lab-Industry Interaction: Various Approaches

For those caught up in the sweep of "metricmania," the pressures increase to find one magic indicator of the value of federal laboratories and their programs. The search is misplaced. The state of the art in R&D evaluation remains primitive, holding no immediate promise of such precision. We believe that the best approach to evaluating R&D impacts is a multimethod approach using multiple indicators, an approach taking advantage of the particular strengths offered by particular measures (Bozeman and Kingsley, in press; Kingsley, Bozeman, and Coker, 1996). In that spirit, this study seeks to examine the value of interactions between federal laboratories and industry bearing in mind the following assumptions:

1. Labs have different strengths and those strengths not only may fail to reinforce one another, they may actually be conflicting.

2. Most technical activities in the laboratories (and indeed, most organizational activity) can be viewed from an opportunity cost perspective (Bozeman and Fellows, 1988; Bozeman and Coker, 1992). That is, it is important to consider not only the value of a given activity but consider its value with respect to possible alternative investments of resources.

3. Many of the benefits of federal laboratory activity are not internalized

to any one firm, and thus an analysis examining one firm's assessment of benefit will typically underreport benefits.

Despite these limitations and the impossibility of coming up with a canonical index of the value of commercial activity, we were able to attack the issue of "return on investment" from a variety of perspectives. One approach is simply to ask for a global indicator of industry participants' satisfaction. While this is certainly not an approach likely to satisfy those interested in precise calculations of value, it is nonetheless an important perceptual variable. People obviously act on the basis of perceptions, and if they perceive that an activity was successful, they are, other things being equal, more likely to continue to engage in that activity. In that sense at least, perceptions matter.

We also sought more "objective" measures. Respondents were asked to estimate the benefits and costs of their participation in the project. The queries about benefits and costs were stated as follows:

> Considering all the possible *benefits* (e.g., training of personnel, developing products or manufacturing processes, receiving technical assistance) but not the costs, what is your estimate of the dollar value (if any) of your company's interaction with the lab on this project?
>
> Regardless of the amount of benefit received by your company from this project, approximately how much does this project *cost* your company in dollar terms? (for example, your company's share of R&D expenses, license fees, salaries for researchers located at the federal lab.)

A related question aimed at deriving a dollar value for the "product" compared the price the company might have had to pay in the private market (in those cases where the product was potentially available from an alternative private source). That question was worded as follows:

> Regardless of the dollar value in terms of benefits derived, approximately how much do you think it would cost your company to purchase the same information (equipment, assistance, technology) from some private source?

One measure of some interest is whether laboratory-industry partnerships lead to the creation (or loss) of jobs. In the past few years, job creation claims have been thrown about wildly by proponents of technology-based economic development programs. Many of the assumptions employed to derive the number of jobs created are highly dubious. However, the job increment-decrement figures presented here appear to have considerably more validity since they are based on no further assumptions than the simple one that industry respondents can provide good information about the gains and losses of positions that result from a given project with which they are, usually, intimately familiar.

The presentation of findings is in four parts. First, before examining the data concerning the benefits of interactions, we provide descriptive information regarding the number and type of interactions that had commercial results. Second, data are presented concerning the perceptual variables related to levels of satisfaction with the interaction. Third, the monetary cost-benefit data are examined. Finally, we consider the effects of the interactions with respect to job creation or loss.

INTERACTIONS LEADING TO COMMERCIAL RESULTS

Figure 6.9 provides information about the commercial status of the interactions examined here. It is important to remember that many of the projects remained in progress at the time of the study and that the data thus do not in all likelihood reflect the ultimate commercial disposition of the projects. At the time industrial respondents were queried, 47 of the interactions had already led to a product being marketed, and in 79 instances products were under development. Perhaps just as important, in 52 instances, existing products had been improved. In only 18 cases had product development efforts failed. In relatively few instances there was either never an intention of developing a product or there were no current plans to develop one.

This profile is interesting in a number of respects. In the first place, it indicates that lab-industry interactions are very much driven by commercial aspirations and, just as important, that those aspirations are often met. Even if one were to assume that only a small percentage of the projects "now developing products" would actually bear fruit, the fact that 18.5% of the projects, many ongoing, had already resulted in a marketed product or service is quite significant. Compared to "productization" rates from intrafirm product development, these figures are impressive.

A question of considerable interest is the difference between interactions resulting in marketed commercial products and services and other interactions. Empirical results from a difference of means F-Test indicated that differences were not numerous but were nonetheless interesting and potentially important. These results are considered below in connection with the questions on which they shed light.

DO CRADAS LEAD TO COMMERCIAL PRODUCTS?

Comparing those interactions that have already led to marketed products or services ("product interactions") to all other interactions ("nonproduct interactions"), one finds a significant relation to the CRADA as a basis of the interaction—CRADAs are considerably more likely not to lead to a product result. Given the emphasis on commercial activity of federal laboratories and

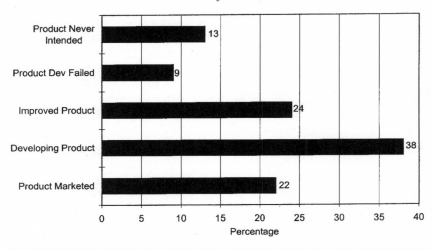

FIGURE 6.9   Product Development Status

on the use of the CRADAs as a vehicle for supporting that activity, one might well wonder what this implies. Since CRADAs have not been in force during the entire period during which this set of interactions has occurred, one might assume it is simply a time-related measure. Such is not the case; there is little relation between the initiation date and commercialization, and few of the earliest projects in the data set led to commercial results.

It is quite possible, though not easy to test, that many of the CRADAs are insubstantial and are more related to push to "increase the CRADA numbers" than to formulate viable, potentially fruitful projects. Another possibility is that the structure of the CRADA itself is not as useful for promoting commercial activity. In particular, provisions constraining federal labs from directly providing funds in support of CRADA projects may limit the effectiveness of this tool.

WHAT TYPES OF COMPANIES DEVELOP COMMERCIAL PRODUCTS?

Many of the companies' attributes included in this data set have no bearings on the likelihood of developing commercial products, but two are especially significant. Tables 6.1 and 6.2 consider differences of means according to company size and age.

The results indicate that the set of companies having marketed products as a result of their interactions with a federal laboratory is both younger and smaller—and substantially so—than the others in the data set. There are several possible explanations for this. Perhaps the most likely is that these newer, smaller companies cannot afford the "luxury" of precommercial or

Table 6.1	Company Size and Product Results from Interactions

Variable	Cases	Mean
Mean Employees		
Product = 0	151	32231
Product = 1	45	4346

$p = .01$ {0 = no, 1 = yes}

Table 6.2	Company Age and Product Results from Interactions

Variable	Cases	Mean
Mean Company Age		
Product = 0	156	44.7
Product = 1	45	26.2

$p = .001$ {0 = no, 1 = yes}

basic research, non-product-related skill development or non-product-related improvement in human capital. It seems more likely that the smaller companies would use the laboratories as a substitute for R&D capacity rather than as a complement. Smaller, newer companies simply have fewer slack resources and operate on a shorter timeline.

Does It Matter Who Initiates the Interaction?

Examination of differences of means indicated that projects were more likely to lead to a commercialized product if the projects were initiated by either the companies' R&D managers or by top managers in the company. Projects developed by bench-level scientists, lab directors or, in most instances, federal laboratory personnel, were no more (or less) likely to lead to commercial results. In part, this may be a size factor. In smaller companies, which we know from the findings above are more likely to commercialize results, the actors are more likely to be in the upper echelons of the hierarchy.

Projects initiated by federal laboratories' technology transfer staff are somewhat more likely to be commercialized. This can probably be explained by the fact that most technology transfer officers have more fully internalized commercial objectives and are intent (as some technical staff may be) to divert resources to curiosity-driven work. Also, the technology transfer staff position often requires that these employees make preliminary market

| Table 6.3 | **Magnitude of Benefits and Results from Interactions** |

Variable	Cases	Mean
$Benefits		
Product = 0	106	$2,042,494
Product = 1	37	$476,000

$p = .01$ {0 = no, 1 = yes}

| Table 6.4 | **Project Cost and Product Results from Interactions** |

Variable	Cases	Mean
$Cost of Interaction		
Product = 0	135	$393,391
Product = 1	40	$536,462

$p = .01$ {0 = no, 1 = yes}

judgments and select for industry interaction those projects with commercial potential. They may also have more experience and acumen in making those judgments.

Do the Companies that Develop Commercial Products Reap More Benefit from Interaction?

The results here are among the more interesting uncovered in this study. One might well conclude that the question of whether companies that develop commercial products benefit more is a question whose answer is so obvious that it is not even worth considering. Not so. The findings clearly suggest that interactions resulting in commercial products are less beneficial than others. The results reported in tables 6.3 and 6.4 are easy enough to interpret and are stark.

The results indicate that companies that did *not* yet develop a commercial product or service reported more than four times the benefits received and only 74% of the cost. Upon reflection, the latter figure is not surprising, as developing products might well be expected to lead to greater cost. But what could explain the strong tendency for those with commercialized products to report *less* benefit? There are several possibilities. Probably the most important explanation is that the federal labs' true value as an industrial partner may be in upstream rather than downstream work. Particularly, basic research and precommercial research may have more value for industrial partners than activities leading directly to product results. This is true because the product-oriented projects are likely more costly (to industry), perhaps less innovative, and may provide little value-added to missions already performed by company R&D. This is not a radical explanation. Indeed, many have argued that there is little reason to expect that increasingly commercial technical activities will lead to enhanced value to industry. Arguably, what is important about the federal lab core competencies is not well reflected in product-oriented research. (This line of reasoning is amplified in the section below dealing with determinants of benefits.)

An obvious point, but certainly one worth making, is that the benefit level in the product development interactions may be lower because the firms involved tend to be smaller ones, presumably with smaller markets and market niches. However, this is perhaps counterbalanced by the fact that the benefit-cost ratios are actually slightly negative ($476K to $536K) among the product development interactions while highly positive, indeed 4 to 1, in the non-product interactions. The only way that the product developing companies can be viewed as having received more benefits is if one divides benefits received by total number of employees. In this case, the benefit is $63 per employee for the non-product companies and $109 per employee for the companies developing products. This seems to be a reach, however, in light of the fact that the benefit-cost ratio is negative. We will return to the question of the benefit and cost accrual, but first turn our attention to more global satisfaction measures.

The "Whole Experience":
Global Assessment of Satisfaction with Lab-Industry Interaction

As a global measure of satisfaction with the company's interaction with the federal laboratory, respondents were asked to agree or disagree with the statement "On balance, working with the federal lab proved to be a good use of our company's time and resources." Figure 6.10 gives the response frequencies. The results indicate that the federal labs' customers are, for the most part, satisfied customers. Indeed, 89% agree or strongly agree that they were satisfied with the interaction, a satisfaction rate that would be reasonably high for almost any business. While one might argue that this is "only" a general perceptual measure, it seems an important one. If the overall experience is perceived as favorable it is more likely to be repeated.

The 11% of customers who were dissatisfied are worth examining in more detail. Differences of means tests were performed by comparing those responding "Disagree" or "Strongly Disagree" on the overall satisfaction with those responding "Agree" or "Strongly Agree." The differences are striking. Clearly, the dissatisfaction is not randomly distributed. The dissatisfied

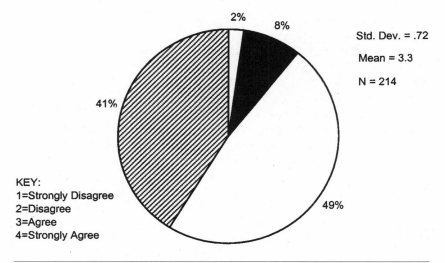

FIGURE 6.10   Customer Satisfaction

respondents are distinctly different. Those respondents reporting dissatisfaction with the federal laboratory interaction tended to have the following characteristics:

Dissatisfaction Associated with

Company scientists at the lab

Lab scientists at the company

Cost of the interaction

Employees hired

Employees fired

Failure in new product development

Previous relationship as a federal lab employee

Previous relationship as a federal lab contractor

Company R&D managers began the project

Satisfaction Associated with

Interaction involved technical assistance

Interaction involved use of specialized federal lab equipment

New product or service developed

Objective included use of lab personnel expertise

Now developing a new product or service

Previous relationship was professional contacts

Previous relationship was informal
Quality of the federal lab important in decision to interact
Lab contacted the company was important in decision to interact
Preexisting personal contacts important in decision to interact.
Federal lab R&D managers initiated the project
The company performed basic research
The lab performed administrative tasks quickly

While each of these findings is suggestive, with only 22 cases of dissatisfaction, it is easy to make too much of them. Thus, let us focus only on what seem to be the overall patterns. In the first place, it seems that projects are more likely to lead to dissatisfaction if they are initiated by the company. The most likely explanation is that companies have particular needs they are seeking to satisfy, and those needs may or may not match lab skills and strengths. Further, and this is highly conjectural, lab scientists and engineers are likely to be more interested and committed to lab-initiated projects because these match best their own interests and R&D agendas.

It is particularly interesting that both new employees hired and employees fired are related to dissatisfaction. Again speculating, we suggest it may be that the new employees hired simply add to the total cost, which, in turn, is strongly related to dissatisfaction (note: benefit levels are not related at all to overall satisfaction).

It seems clear that the motives for working with the lab are an important factor in satisfaction. If the company comes to the lab because of the special skills of lab employees or special equipment of the lab, the experience is satisfying. However, if the company views the lab as a "generic problem solver" there is less likelihood of satisfaction.

JOB CREATION

Another possible benefit derived from companies' interaction with labs is job creation. While this is not necessarily an unmixed benefit for the company (the company might, in fact, benefit from the reduction of employees through the substitution of technology or lab-based labor), it is important from the standpoint of social benefit derived from the labs. On average, 1.6 new jobs were created by these interactions (which must be weighed against the .05 jobs displaced as a result of the interactions). Fully 90% of the projects do not result in new hires. Of all the projects examined, only 3 resulted in hires of more than 10 new employees. One project resulting in 200 new hires vastly skewed the results. Thus, it is probably safe to conclude that the vast majority of lab-company interactions do not result, at least not directly, in any job creation or loss.

ESTIMATES OF MONETARY VALUE FROM INTERACTIONS:
COSTS, BENEFITS, AND BENEFIT-COST RATIOS

While the reported level of satisfaction gives us some insight into interactions between federal labs and industry and their success, it is also useful to seek more precise indices. This has been done by having respondents indicate the monetary value of the interactions to their company. The results are encouraging. The average level of benefit reported was $1,548,073. The average level of cost was $448,765. While it is tempting to say that the companies had a 4:1 multiplier for their interactions with the federal laboratories, it must be understood that in many instances the companies have essentially no cost for the interaction. Looking only at those companies that sustained costs for the interaction, the mean rate of return was 2.67. That is, they valued the benefits at a rate of 2.67 greater than the amount expended as company cost. This is still quite a good investment. However, this figure masks a good deal of diversity. A significant percentage (26.6%) of the companies did not value the benefits as highly as the cost or, in effect, did not break even in their investment. Figure 6.11 provides the distribution of benefits and shows that they are exceedingly varied. Costs are even more varied and, because of extreme outliers, are not easily graphed. It is also notable that a good number of respondents, a total of 23, report no benefit from the interaction.

After the investigation of factors relating to the overall satisfaction with the interaction, it is useful to consider the relationship of satisfaction to monetary cost and benefits. The results are not unexpected. The proportion of benefits to costs goes up unsurprisingly with satisfaction. It is interesting,

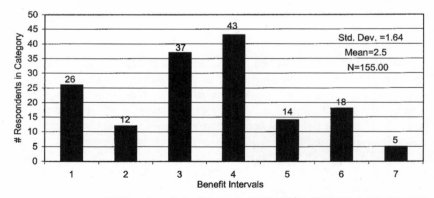

KEY:0=no benefit; 1=<$10,000; 2=<$100,000; 3=<$500,000; 6=$10,000,000+

FIGURE 6.11   Distribution of Benefits

however, that even for those projects where the respondent indicated "disagree," working with the lab was a good use of time and resources. The benefits still exceed the costs.

In most instances (70%), the respondents indicated that the output provided by the federal laboratory was not available from another source. This in itself is perhaps a useful indicator of the value of the output. However, for 83 of the projects examined the respondents reported that the output could have been obtained from another source. The mean value in those cases where there were alternative markets is reported as $513,543. If we consider that the mean cost reported for companies' interaction with the lab is $417,010, this seems, again, a good return. More to the point, a ratio of the cost to value of purchased information is conservative since many of those reporting a cost did not report an alternative market value because of the unique nature of the output.

The primary focus here is on the ratio of costs to benefits. Since it is difficult to analyze data with "0" in the numerator or the denominator, the focus is on those projects that reported costs and benefits greater than zero. Figure 6.12 shows the distribution of benefit-cost ratios. The tall bar in the middle is the breakeven point; any number greater than one is a positive benefit-cost interaction.

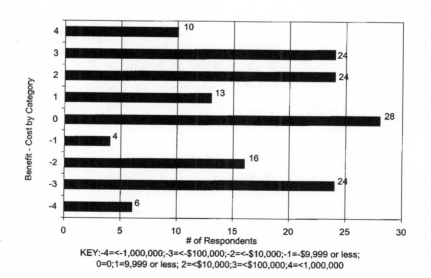

KEY:-4=<-1,000,000;-3=<-$100,000;-2=<-$10,000;-1=-$9,999 or less;
0=0;1=9,999 or less; 2=<$10,000;3=<$100,000;4=<1,000,000

FIGURE 6.12 Benefit-Cost Difference by Category

From the Standpoint of Benefits and Costs, What Factors Relate to Project Success?

There are 34 cases where the company invested its own money (i.e., had some costs) and had benefits of at least twice the costs. Focusing on those clear-cut successes, we can try to determine how these success stories differed from other apparently less successful interactions. There are also 27 projects that report some benefits, ranging from a few thousand dollars to millions, but no cost. These projects are certainly important for benefit counting, but it is not as clear how they should be analyzed from a cost-benefit perspective.

A difference of means F-test was performed. It was based on a comparison of those interactions that had a better than 2 to 1 ratio of benefits to cost with those that had some cost and benefit but a lesser return rate. The "successes" had some characteristics that would be expected by virtue of the index. For example, projects that resulted in a commercial product were much more likely to be in the "more successful" category, as were projects with a high monetary value for purchase information or resources in alternative markets. Similarly, projects that tried to develop commercial products but failed were much more likely to be in the "less successful" category. But other findings were not quite so intuitive.

"Double-Your-Money" Interactions Have the Following Characteristics:

Smaller size companies

CRADAs

Not licensed technology

Do not involve exchange of scientists and engineers

Do not involve decrease in company personnel

Lead to improvements in company scientists' and engineers' skills

Company personnel former lab employees

Company personnel former lab consultants

Prior interaction with lab not a decision criterion

Previous project with lab not a decision criterion

Lab did not perform precommercial R&D

Lab performed development

Company not in existence for a long time

High proportion of R&D personnel to total personnel

In sum, the profile of a company with interactions having a high ratio of benefits to costs is, from the company standpoint, a smaller, high-tech firm that is not old and established. The participants are likely to have worked at the lab or to have been consultants. But the lab was chosen as an R&D partner not because of previous projects but because the lab offered special skills and resources. The interaction was likely to have been based on a CRADA and was not likely to have involved the licensing of technology. The lab was likely to have performed development work but not precommercial R&D.

It is important to note that despite the seemingly more "tangible" indicator of cost-benefit ratios, the measure is an incomplete indicator of "success." For example, it is clearly the case that the value of precommercial R&D and basic research is likely to be underestimated by such direct monetary indicators, and the value of development is likely to be overestimated. The reasoning is simple—if immediate return on investment is the issue, precommercial and basic research are, by their very nature, unlikely to show a positive contribution. In fact, in many ways the global measure of success, which permits the respondent to reflect upon a variety of criteria, albeit less precisely, is a more useful index. Nevertheless, it is useful to know that interaction with the labs is in many cases a very good "investment," even by the standard of short-term returns. Here, it is worth reiterating our earlier point: any evaluation of commercial impacts of the federal labs based only on one perspective or one set of indicators is likely to lead one down the wrong path.

During the past two decades several public policy initiatives have sought to enhance the value of federal laboratories as commercial partners with U.S. industry. By this point, there have been thousands of commercial interactions between federal laboratories and private firms. Our results, based on data collected for 229 projects, indicate that, on average, the interactions are assessed as quite positive, as having considerable net benefit to the firms. But there is considerable variance for most of the benefit-cost indicators.

A fundamental issue is the incidence of commercial outcomes. For the set of interactions examined here, *22% had already led to a new product, process, or service being marketed—a high rate given the fact that most of the projects began after 1990.* In 38% of the cases, product development was underway at the time the data were reported. In 24% of the cases an existing product had been improved. In comparing those companies who had already developed projects to all other companies, the ones with products developed from the interactions had the following characteristics: (1) smaller than the average for all companies in the data base (12,000 employees on average v. 25,000 for the sample), (2) high levels of R&D intensity (R&D employees as a percentage of total employees), (3) established more recently (average 27 years, compared to 45 years for all firms in the database); (4) surprisingly, report *less* absolute benefit than the average for all firms. This latter point is perhaps explained by the fact that the firms are, on average, a good deal smaller and, thus, the benefit level is reduced by the size of the firm and the interaction. A number of characteristics of the interaction appear related to whether it has been commercialized. Those projects initiated by technology transfer staff or top management of the lab and of the company are more likely to have been commercialized.

From the standpoint of overall satisfaction with the project, *89% agreed that the interaction was a good use of their company's resources.* Even many (41%) who reported that costs exceeded benefits nonetheless had a high level of "customer satisfaction." Those who are *not* satisfied with their interaction with the federal lab (23 cases, 11%) are more likely to have been involved in an interaction in which a license was purchased, are more likely to have had reduced personnel as a result of the interaction, and are more likely to have tried to market a product but failed.

The one criterion by which the laboratory-industry interactions could not be said to have been *particularly* successful is job creation. *More than 90% of the projects did not result in a single new hire.* We underscore that our method involved calculating only new jobs at the participating company, not multiplier effects to, for example, suppliers and vendors. It must be remembered that many of the projects are of recent vintage. But the fact that so many projects have already led to the introduction of products while so few have led to new hires seems to provide some evidence that the failure to produce employment results is not simply due to insufficient time. Indeed, the pre-1985 projects had a job creation rate inferior to the post-1990 ones.

The interpretation of the costs and benefit numbers provided by the respondents requires much care. A general finding is that by virtually any of the measures, the lab-industry interactions are, *in aggregate*, a "good investment." But this does not mean that every interaction pays off handsomely for the firm. There are vast differences in outcomes, with some firms having huge payoffs and others suffering substantial losses. The average level of benefit reported was $1,548,03. The average level of cost was $448,765. But in the absence of additional information, these figures are not particularly informative. In the first place, they mask a great deal of variance. In the second place, many companies (43%) report no costs and several (17%) report no benefit. *Focusing only on the firms that sustained both costs and benefits, then the average benefit is $1.8 million and the average cost is more than $544,000, a return of more than three to one.* If we add those companies that reported zero benefit, then the ration falls to 2.8:1.

We provide a number of approaches for comparing benefits and costs, but one of the most straightforward and conceptually meaningful is to simply subtract costs from benefits to determine the net benefit. Even in those cases where the number is negative, it is a valid negative. *The average (arithmetic mean) net benefit for all projects is $1,087,500.* But this number is skewed by a few "big winner" projects. Indeed, projects are only somewhat more likely to be net earners (49%) than to break even (18%) or have net losses (33%).

Determining the factors related to "success" is complicated in several ways.

Our results make clear that the subjective assessment of value (in terms of overall satisfaction with the interaction) often yields results quite different from estimates of monetary costs and benefits. This is not to say that one approach is more valid than another—"customer satisfaction" and economic cost-benefit are both important. In the case of the cost-benefit data, quite different results accrue from, among other possibilities, including or excluding real zeros, measuring first differences, measuring cost-benefit ratios, and adjusting or not adjusting for outliers on the distribution. Again, each of these indicators is valid; they simply offer different information.

A few factors seem to relate to virtually all the evaluative indicators. One of the most important factors has to do with whether there is a product marketed from an interaction (or whether the development process is underway). The development of a product is highly salutary with respect to all measures of benefit—overall satisfaction, having a positive return on investment, and having a large net benefit. Similarly, a failed attempt to develop a product has the effect of depressing all measures of benefit. Two other factors having across-the-board importance: if the lab contacts the company and if the company has previous experience working with the lab the interaction is likely to have greater value.

These findings have some relevance to the current storm of activity surrounding the possible realignment and refocusing of the federal laboratory system:

- In aggregate, the interactions between federal labs and industry do appear to create significant economic value and, from the standpoint of the firms involved, receive a quite positive assessment.
- There is tremendous variation in the economic value of the interactions.
- Industry-lab interactions do *not* (as often advertised) create many new jobs.
- Particular characteristics of the company and the interaction relate to the likelihood of success, implying that empirical results such as ours can be used to develop strategies for increasing the benefit of lab-company interactions.

One of the strategic questions suggested by the results is whether to develop policies to help grow small business or whether to pursue "big winners." Just a few of the projects have such enormous benefit (in excess of $10,000,000) as to provide an excellent justification for lab technology commercialization policies. But there are good arguments for the labs serving as a spur to small business, perhaps providing less absolute benefit but a great deal of "bang for the buck," especially in terms of small companies' long-range future. There is, of course, no obvious reason why both strategies should not be continued, but they may involve very different evaluation approaches, different implementation approaches, and even targeting of different labs.

Another important strategic issue is the role of basic research in commercialization. It is easy to argue that the labs, many of which have strong basic research capabilities, should play a role as basic research provider (or partner) to companies. But it is also plausible that commercialization efforts are more likely to pay off when they are centered on product development. Actually, our results seem to support both views. Activities focused on product development are likely to provide substantial benefits. Basic research does not enhance the likelihood of commercial payoff from interactions. But those projects that are "big winners," the ones with huge returns, are disproportionately basic research projects—indicating that just as basic research itself is a high-risk, high-payoff activity, basic research aimed at commercialization seems to have the same character.

Finally, while we learned a good deal from this phase of the NCRDP about the value of interactions between federal laboratories and industry, many pieces of the puzzle are still missing. In the first place, this is a study of companies that have interacted with federal laboratories, and it would be useful to examine some that have not done so. To what extent is the lack of interest in working with federal labs due to self-conscious avoidance decisions and to what extent is it due to lack of exposure to the labs or to other such nonsystematic factors? Second, what are the opportunity costs of laboratories' involvement in commercial activity? When labs are working with companies there are other activities that could be pursued but are not. What is the value of industry-federal lab interaction *compared to* other lab activities? Any comprehensive analysis of the "competitiveness potential" of the laboratories will need to take such factors into account.

The bottom line for this preliminary assessment of the cooperative technology paradigm is that partnerships between federal labs and industry *can* work and often *do* work, but their success is far from being uniform.

CHAPTER SEVEN

Redesigning Federal Laboratories

Twenty Years of Study Panels and
Their Recommendations

With $25 billion on the table annually, and with facilities the size of small cities, it is little wonder that the U.S. has struggled with what to do with its federal laboratories. For example, how do you direct the future of labs devoted for 50 years to the making of weapons of mass destruction? Since 1978, more than 20 major commissions or task force groups (see appendix 2 for complete list) have been chartered to help address some version of this question. Chartered by the president, secretary of defense, secretary of energy, congress or others, each group has posed essentially the same question: How can the various federal laboratories be enhanced to be of greater value to national purpose and, especially, economic development? The targets in these reviews have often been the major multiprogram labs of the Department of Energy and the labs of the Department of Defense. Given both the magnitude of activity in the leading DOE and DOD labs and the changes in national priority associated with their historic mission, it is little wonder that there exists concern about the deployment of this multibillion dollar asset and its multibillion dollar annual budget.

The findings of these various commissions and task forces were based on many of the assumptions associated with the general movement toward the cooperative research paradigm that were detailed in earlier chapters. These findings were also based on anecdotal evidence about how to increase the effectiveness and efficiency of the federal laboratories. A summary of the findings of several representative efforts is presented below. In each case, we assess the findings and recommendations of the panels, in many cases relying on our institutional design framework and NCRDP research findings.

1993 Council on Competitiveness: "Industry as a Customer of the Federal Laboratories"

With the election of Bill Clinton as president in 1992 and with the development of a new national technology policy as a stated objective of the new administration, a number of groups stepped up to develop their concepts. Prominent among these groups was the blue-ribbon panel established by the Council on Competitiveness. The council was established in the early 1990s as a forum and interest group for those companies and universities interested in the development of policies intended to maintain national technological competitiveness. While pursuing a number of issues and working Capitol Hill on a number of fronts, the council took it upon itself to attempt to deal with what was a major question of the day, namely, what to do with the federal labs. Dominated by industrial representation, the council formed a working group focused on designing a future role for federal labs. This working group and ultimately the entire council reported a number of specific recommendations on how to enhance the technology transfer process between government and industry. These recommendations were built on the assumption that federal R&D labs were untapped, underutilized sources of new technology. Further, it was assumed that there existed a need for clear action intended to "make the labs more responsive to industry's needs (i.e., more customer-driven)." This basic assumption is derived from the belief that the labs are a dormant source of new technology and that all that is needed to enhance U.S. competitiveness is to provide the mechanisms for the labs to release this latent source of technology. In that context the council recommended the following:

- Congress and the executive agencies should assign 10% of the budget of DOE and NASA labs to technology transfer programs.

There is a general assumption on which this recommendation is based, namely, that there is insufficient spending on technology transfer and that this is the key factor in linking with industry. There is no evidence that lack of government funding is the problem. Some labs as a result of size, breadth of mission, and culture are equipped to work with industry in the technology transfer process; others are not. Some laboratories' basic science mission limits their ability to contribute to a technology transfer program. They make contributions to the basic stock of knowledge, which can have dramatic effects on overall technology patterns but have less potential for one-to-one technology transfer.

It is an aggravating problem in most federal labs, that they are not funded as units. Most federal labs are funded via individual program managers who are attempting to attain various mission objectives through research groups located at the various labs. Outside of the USDA, there is little meaningful institutional-level planning for labs in agencies with multiple labs under their jurisdiction. Therefore, the result is that even with a 10% commitment to directors, systematic technology transfer will remain problematic.

Most important, many groups within federal labs have nothing significant to offer as technology seeds or development opportunities. Labs focusing predominantly on either fundamental research results, technical service in support of industry, technical service in support of the government, or classified military research generally offer little commercial technology. Indeed, what some of these labs offer is more important for the NIS than a set of commercial technology opportunities. Assigning 10% of all lab budgets for technology transfer programs would, without discretion, weaken or destroy programs of the type listed above. A universal technology transfer mandate fails to recognize the variation and uniqueness that exists in the federal R&D lab community. Let us remember the comparative advantage principle.

- Increase laboratory directors' discretionary spending to 5–10% of their annual budgets

We know that laboratory directors are positively influenced by an economic development mission. We also know that the lack of flexible dollars is a critical weakness in the design of federal labs. If we assume that each laboratory would be defined around a specific mission and that this specific mission statement included a definition of its role within industry, then a federal laboratory could, with such discretionary funds, potentially increase its effectiveness with industry significantly. Often, a laboratory works in realms where close collaboration with industry is desired as a precursor to successful technology transfer. As indicated by our research, this is an important factor. In this situation, lab directors might use such discretionary funding to cofund centers for joint technology development and transfer. Such center settings have proven most effective in stimulating traditional technology transfer.

At the same time, it must be remembered that many federal labs have little technology to offer industry. What they have is basic knowledge, technical insight, and know-how. In these cases, such discretionary dollars could be used to build enhanced two-way information linkages with industry and to tackle small know-how problems for small and medium-sized companies. This type of knowledge transfer, as evidenced in the findings in chapter 5, can be of tremendous value to industry, but it would not fall under the general rubric of technology transfer. In fact, because labs generally lack organizations connecting their useful knowledge and industry, the transaction cost for acquiring such knowledge is so high that our findings show that industry views most information of this type as a negative input to the technology development process. Discretionary spending could be used to establish a bridge between the labs and industry in industry-relevant ways.

- National technology infrastructure projects should be launched so as to strengthen U.S. industrial competitiveness and foster cooperation between industry and federal labs

Here it is assumed that critical areas such as telecommunications or the environment can be identified as national projects and that federal labs can be directed to cooperate by focusing their energies on these national projects.

Our empirical evidence suggests that the diversity among federal labs, in terms of mission, skills, and design, is so great that any across-the-board effort to link these labs with national projects will be problematic (Crow and Bozeman, 1991; Bozeman and Crow, 1995). Our findings regarding industry's evaluation of its technology transfer interactions with federal labs indicate that working with industry is not a simple process of redirecting a lab's energies (Bozeman, Papadakis, and Coker, 1995). A key factor is whether or not the research group based at the federal laboratory has anything to offer, either knowledge or technology, to a market-driven company trying to develop or improve a product or process.

We must also remember that evidence suggests that some labs are equipped to interact with industry very effectively, while others, because of their focus on the long term, their particular culture, or their problem set, should be viewed as building blocks for national technology projects (Bozeman, 1994a; Bozeman and Coker, 1992). There is a need for a highly detailed mission definition and the building of a team of labs that link basic and applied research capability in areas related to particular national technology infrastructure projects.

- Federal laboratory directors should have full legal authority to negotiate, sign, and execute cooperative R&D ventures with industry.

Our research indicates that not only is government red tape significant in general, but the red tape within federal labs stands out (Bozeman, Reed, and Scott, 1992; Crow and Bozeman, 1989a). Our findings indicate that all decisions are slower in general and that the release of research findings is, except for basic research, slower than in all other lab types (Crow and Bozeman, 1991). Therefore, granting lab directors authority to enter into such agreements without providing for some level of overall autonomy for the lab would not result in a great improvement with regard to cooperative R&D ventures.

- DOD should establish an outreach program to make R&D and technical expertise in DOD labs more accessible to civilian industry.

The development and use of bridging organizations for the purpose of industry outreach and as a means of cooperating with industry has proven to be a particularly successful arena for activities related to technology transfer (Crow and Bozeman, 1989a). It is likely that research centers organized to act as a bridge between industry and DOD labs could prove effective as transfer enhancement tools.

- Federal laboratories (non-DOD) should establish industrial advisory committees to assist in the planning and execution of programs related to technology transfer.

Many labs have already established industrial advisory groups. Our findings indicate that such groups, if organized into collectives with a research center orientation, could have a positive effect on the process of technology transfer (Bozeman and Crow, 1991a; Bozeman and Coker, 1992). While we did not look at these advisory groups as committees, we did view them as a form of enhanced research coordination offering immense value from the industrial perspective.

1978 The Multiprogram Laboratories (A Report of the DOE before the Current Laboratory Reevaluation): A National Resource for Nonnuclear Energy Research, Development, and Demonstration

Shortly before intense reevaluation of federal labs got underway in the early 1980s, Congress inquired of the DOE to define how the national labs under its jurisdiction could be harnessed toward the development of nonnuclear energy technologies. This request came in the first full year of the operation of the newly formed Department of Energy and was the result of a number of concerns about how the old labs of the Atomic Energy Commission could best be applied in the context of a civilian energy development agency. The questions of the day evolved around the general concern of how to increase the return to the government from its investment, but also considered defense and civilian deployment of lab capacity. This precursor analysis and set of recommendations from nearly 20 years ago asked the same essential questions that are being asked today. How can federal laboratories be redirected toward new national missions and purposes? Twenty years ago the U.S. established a national priority, nonnuclear energy technology development. The nation attempted to focus its most significant federal lab asset, the DOE multiprogram labs, on that problem. Much like the Council on Competitiveness' recommendation of 1993, the thought was that nonnuclear energy technology would spur cooperation and technological success. The specific 1978 recommendations included:

- Organize the national labs for national focus rather than programmatic focus.

The thought here is that the labs are controlled by the various DOE program offices, such as defense or nuclear programs or basic energy science programs, and that this mission control limits the flexibility of the deployment of the asset. Our research indicates that the most successful labs are larger and have multiple missions (Bozeman and Crow, 1988, 1995). Thus, it seems sensible to raise the labs to a higher organizational level and detrimental to narrow the mission of the labs.

This historic recommendation reveals also that in the years since 1978 the status of the labs at DOE has not changed. Our findings from a detailed 1984 analysis of DOE labs indicates that DOE laboratories were not significantly changed between 1978 and 1984 (Crow, 1985; Crow and Bozeman, 1987b). Our later work between 1984 and 1996 confirmed once again that this basic design feature (program-level control) has been maintained and has, generally, had negative consequences (Bozeman and Crow, 1992, 1995).

- Develop coordinated laboratory missions maximizing individual and lab group competencies

In the 1980s and 1990s the federal R&D laboratories demonstrated a high level of cooperation and linkage among themselves. This is less the case among the DOE labs discussed circa 1978.

Interestingly, the 1978 report by the GAO was clearly based on the market failure model for federal labs, finding that it was proper for federal labs to move in when and where there are insufficient incentives for private sector research. But in all of the recommendations there was only one mention of industry. Even in this one instance, industry was mentioned only as a subcontractor helping to fulfill national energy research objectives in coordination with the national labs.

1994 Changes and Challenges at the DOE Laboratories (The Galvin Commission)

Two years into the Clinton administration the issue of what to do with the national labs remained unresolved (as it still does). Just the DOE labs alone were costing billions of dollars per year, and their energy and defense missions were both being questioned. The secretary of energy turned to Robert Galvin, the former chairman of Motorola, and an industry-dominated panel to map out some alternative futures for the DOE national labs. The report, which got broad exposure through the press, concluded that fundamental changes were required. The report recommended that the government carry out of a number of steps intended to enhance performance and return on national investment. These steps include the following:

- Develop new management systems based on quality principles

Most federal R&D labs already have strong notions about quality. When asked to rank various labs or to list the highest quality research labs, lab directors have little difficulty responding. Regarding new management, we do know that R&D labs are particularly sensitive to external instability and political and managerial uncertainty. New management system designs must consider the differences among the labs, the comparative advantage principle. Moreover, the organizational structure and internal management of the labs have much less impact on performance than external factors affecting laboratory management. Yes, management changes are badly needed, but at levels outside the laboratory.

- Develop mission to emphasize the use of federal lab core competencies as a national resource in collaboration with the private sector and other federal agencies.

Complex federal laboratory missions have evolved, in most cases, over several decades. Our research indicates that R&D labs have evolved to perform particular, although often broad missions (Bozeman and Crow, 1990). Public science labs focus on basic science and precompetitive applied research with little direct government intervention into their particular research agendas. These public science labs have evolved into organizations whose effectiveness could be inhibited through the introduction of market factors. One might wish to have public science labs coordinate with industry in carefully defined ways. Collaboration might be limited to enhanced information flow to targeted industries, industrial-government scientist exchange, bridging centers that build on basic research findings, and enhanced patenting and licensing efforts.

On the other hand, a public technology lab could serve as a key technology developer for enabling technologies in a particular sector. For instance, Agricultural Research Service labs involved in new product development from agricultural feedstocks could be clustered in collaborative teams with industry based on a research agenda developed in conjunction with industry. In this case, the federal labs could tackle those elements of a technology development problem that benefit all private developers or those problems that the federal lab is best equipped to address on its own.

- Focus laboratories on long-term R&D as a supplement to address the decline in American industrial investment in long-term R&D

Laboratories already have foci that are difficult to change. Early results from the NCRDP indicate that efforts to alter missions often have net negative effects (Bozeman and Crow, 1990; Crow and Bozeman, 1991). Many federal laboratories are already focused on long-term R&D. Enhancing their linkages to support U.S. industry would require whole new strategies for R&D project formulation and agenda-setting.

At present, the dominant forces affecting federal R&D labs and their agenda-setting process include:

- Federal appropriations
- Missions defined by the parent agency
- Political environment of parent agency mission
- Degree of laboratory autonomy

Laboratory missions have typically been focused on service to the parent agency. Unless there is dramatic change in the core mission of the agencies themselves, the laboratories have limited ability to independently alter long-standing research patterns.

- Expand all laboratory missions to provide industrial, educational resource and environmental assistance to the region in which it is located.

Our findings indicate that labs already perform this type of mission to a significant extent, but they receive few resources for such activity (Bozeman, Papadakis, and Coker, 1995). Present barriers to expansion are bureaucratic and financial. This is particularly true with extensive technical service activities. Increased technical assistance will, for most federal laboratories (the great majority of which now perform only limited technical assistance), lead to reduced overall effectiveness. The comparative advantage principle tells us that only selected laboratories should have significant technical assistance roles.

- Develop more partnerships with industry

Industrially active technology development organizations in the U.S. are typically small emerging science or engineering groups. These organizations have few if any linkages with federal labs. On a larger corporate scale, the research products from federal labs are perceived, at best, as a neutral and often as a negative factor in the process of developing corporate technology strategies (Bozeman and Crow, 1991a). Simply building more partnerships is at best problematic and at worst not likely to occur without substantially rethinking how small and medium-sized companies interact with the labs. It is not the *number* of partnerships that is important but their *quality* (Bozeman, Papadakis and Coker, 1995b).

1993 Defense Conversion: Redirecting R&D, Office of Technology Assessment

Although eliminated in 1996, the Office of Technology Assessment (OTA) for nearly 25 years had an active policy analysis function in the area of science and technology policy. In 1993 OTA was asked to review the issues surrounding R&D as affected by the effort at the time to link military research more closely with civilian needs, an effort known as "defense conversion." At the time Congress was interested in the overall issue of how best to link the attainment of military objectives with the possible secondary benefits to the civilian sector. The assumption was that investments in military R&D, which at the time totaled more than 55% of the entire federal R&D investment, might through careful planning be parlayed into specific civilian applications for industry. As a result of this general thrust a number of new programs where developed. Although shortened by the 104th Congress in 1995, this effort did involve a rethinking of the design of defense R&D labs. Specific OTA recommendations included:

- Speed up cooperative R&D agreements (CRADAs)

Here it is assumed that CRADAs are the means by which technology transfer can be best accomplished. However, our research indicates that CRADAs are not always the best path to successful technology transfer (Bozeman and Coker, 1992; Bozeman, Papadakis, and Coker, 1995). CRADAs came to be little more than a popular symbol for an incoherent jumble of laboratory activities. By becoming everything, they became nothing—at least nothing distinct. This OTA recommendation focuses on numerical assessments of one ambiguous technology transfer vehicle.

- Enhance local autonomy of labs

Again, the OTA focus is on CRADAs. This focus fails to recognize those areas in which labs either address market failure research needs or focus on research transfer where CRADAs would be inappropriate. The emphasis of these recommendations reflects a narrow nonsystem approach to thinking about the labs. At the same time, a more general recommendation for increased local autonomy matches well with our findings that laboratories that set their own technical agenda are the most effective.

- Create a lab rationalization commission

According to this recommendation, a commission would be established to determine which DOE & DOD labs remain as critical assets of the nation. Such a commission would necessarily need to consider the design and structure of all federal labs and not just the DOE/DOD laboratory community. Focusing such efforts only on DOD and DOE labs would fail to recognize what our findings reveal, namely, that the government R&D laboratory community and the university community are highly interwoven (Bozeman and Crow, 1990, 1991a). We also know that within individual labs, particularly large federal labs, there are multiple missions and multiple linkages. Therefore, a lab-scale review would be unable to accurately analyze the scope of lab variability and interorganizational relations. Such a commission would need to map and understand the complex interorganizational relationships among federal laboratories as well as those between federal laboratories and other laboratories. An effective assessment must consider more than DOD and DOE laboratories alone.

- Establish stronger incentives for technology transfer.

Our findings (Bozeman, 1994a; Bozeman and Coker, 1992) indicate that the strongest incentives for laboratory involvement in technology transfer are:

- economic development mission
- collaborative centers
- new national missions

It is likely that the behavior of federal labs will be altered by the improvement of institutional incentives, flexible dollars, 'lab autonomy, royalty income return, and personal incentives. Consider the case of universities. Since 1981 universities have been marketing their patents on an exclusive basis with substantial economic incentives. As a result of these efforts, the patents issued to the universities in 1992 alone equaled the total of the previous ten years. Licensing income and therefore new product sales have shown the same responsiveness.

1983 Report of the White House Science Council: Federal Laboratory Review Panel

In 1982 and 1983, in the middle of the first term of the Reagan administration, the economic stability of the nation seemed uncertain. While the economic downturn of the period was a function of a number of factors, it was well understood that one issue of importance to national economic health was the structure and performance of the NIS. There was recognition of the need for greater attention to the federal R&D investment and the U.S. R&D system. At the same time there was a massive increase in spending related to national defense, a major factor in the doubling of the national debt between 1981 and 1989. The process of increasing this spending in 1981–1985 put tremendous stress on the rest of the federal budget forcing new scrutiny of the expensive, uncertain investments in federal R&D labs. This complex set of motivations led to the development of a White House panel on the future of the national labs.

This 1983 report, known popularly as the Packard Report (after the panel chair, David Packard), was perhaps the most profound and far-reaching of all of the studies of the 1980s and 1990s. It took the first radical departure from the party line (science is inherently good, labs are national treasures) in the then 40-year history of major investments in federal research. The report was a scathing attack on the federal research laboratory establishment. The Packard report's recommendations included:

- Laboratories should be given specific missions from which their performance could be measured

In many cases, particularly among the larger federal labs, the laboratory is more a community of loosely attached research units than a tightly coordinated, centrally managed organization. For example, at Oak Ridge National Laboratory, there are groups producing critical materials for DOE Defense Programs, new materials being developed for DOE/Basic Energy Sciences, and social science studies funded by DOE and other agencies. Each of these groups follows its Washington-based program leaders, and on occasion they are subject to the administrative leadership of the laboratory director. Our research has not found federal laboratories to be missionless, but "mission-full" (Bozeman and Crow, 1995). Multiple missions, sometimes conflicting, are the norm at the larger laboratories.

Among smaller federal labs there are much more specific and focused missions. The National Animal Disease Center of the USDA is an example of a smaller lab that clearly knows its mission. The Naval Civil Engineering Laboratory has no confusion about its mission. What many of these labs do not know is whether or not their mission also includes diverting time, energy, and resources toward working with industry. Individual laboratories with self-defined research agendas fitting clear missions do appear to be the most effective.

- Size of federal laboratories should be determined by their missions and quality

Both mission diversity and research niche are the principal determinants of size. It is important to understand, in addition, that the research component of some federal labs is smaller than the technical support, material manufacturing, and service components. A significant difference exists between a research group using parallel processors to develop new algorithms for the purpose of solving geophysical problems and a group using the same computer technology to optimize a tactical nuclear counterattack scenario. Yes, size should be determined by mission and quality. Missions, however, must first be carefully understood.

- Develop a scientific/technical personnel system independent of civil service

All laboratories operating outside of the civil service have lower levels of red tape and, as a result, respond more quickly and are quicker in almost every way. They can hire or fire, acquire new equipment, change structure, and move into new directions more rapidly and with greater ease (Crow and Bozeman, 1989a; Bozeman and Crow, 1991a). Dependence on the civil service structure is seen by a majority of federal lab directors as a barrier to successful science and successful technology transfer (Bozeman and Crow, 1988, 1995).

- Federal laboratories should open up their facilities to industry and universities

In general, facility use by industry remains a minor factor in the success of technology development or technology transfer (Bozeman, 1994a). There are, of course, unique facilities that have, for the most part, been open and available since the early 1980s. Some of these facilities are important in advancing industrial technology, but many more are important in advancing fundamental science.

- R&D interactions between federal laboratories and industry should be increased by more exchange of knowledge and personnel as collaborative projects and by industry funding of laboratory work.

Our evidence indicates that rapid movement toward this recommendation could lead to a significant decrease in effectiveness, high levels of instability, and a general decline in the overall quality of federal labs so affected (Bozeman and Crow, 1995). However, those having a long history of working closely with industry and, especially, those designed for this type of mission are an exception. If there is an interest in enhancing laboratory-industry collaboration, a better approach is to create laboratories with this specific mission, rather than appending it, as an afterthought, to long-standing, successful missions.

During the past two decades, major reports and task force statements concerning federal labs have been issued at a rate exceeding one per year. Without exception, the principal recommendations of these key studies are weakened by the fact that there is insufficient recognition of the system-level variables and institutional factors that govern laboratory behavior. As a result, the implementation of many of these recommendations (including some already being undertaken) could lead to a diminution of the contribution of the 700 federal laboratories operating the U.S. NSI and an inefficient match of the federal labs to the other labs among the 16,000.

CHAPTER EIGHT

Strategic Analysis and Design Recommendations for Federal Laboratories in the National Innovation System

Having provided an analysis of the roles of federal laboratories in a previous chapter, along with recommendations about how those roles pertain to policy issues facing the federal labs, we consider in this chapter some of the elements of strategic science and technology policy and recommend some specific strategic approaches. In particular we consider here the federal labs within the NIS and their relation to other labs in the NIS.

The entire R&D laboratory community is influenced by public policy tools such as direct R&D funding to more than 40% of all labs in the country, R&D tax credits, and Small Business Innovation Research (SBIR) funding. Yet, the national science and technology debate of the late 1990s has been focused on the future of federal R&D laboratories, one important element of the NIS. We believe that this debate should be institutional in its character. We employ premises of institutional design in presenting our recommendations.

One of the key questions here is "how does one develop strategies for deploying the federal laboratory system in the public interest?" We have three basic answers to this question. First, there is a need for a great deal more strategy-relevant information. Second, the "new ways of thinking about the NIS," such as developed in the taxonomy detailed in chapter 4, can provide new tools for analysts. Third, the relationship between technical mission and agency or department affiliation must be reexamined.

Strategy-Relevant Information

Given the current interest in redesigning and realigning federal laboratories, the scarcity of information available for these strategic decisions is remarkable. There is no shortage of opinions, advice, and stories, but the accumulated *relevant* data is minimal. Given that some federal lab directors feel

overwhelmed by the number of inquires for information, especially questionnaires, but other types of data as well, we might ask how there can be a
clear information deficit and, at the same time, an information-gathering
burden. The answers to this conundrum are simple—too much of the wrong
sort of information is being gathered from, essentially, the same small group
of people. The directors of the largest federal laboratories (and, for that matter, the largest industrial laboratories) are beset with multiple requests for
information. Perhaps most under siege are the DOE multiprogram laboratories, which are in the "unfortunate" situation (from a data target perspective) of being large in personnel and budgets, devoted to missions that are in
decline or flux (high energy physics, weapons development/nuclear material
containment), potentially relevant to commercial objectives, and highly visible to elected officials. However, there is little or no coordination among
the many private and government seekers of information.

In our view, major redesign of the federal laboratory system in the absence
of strategic information about the NIS and the R&D laboratory community
is a significant impending threat to the well-being of the U.S. technical enterprise. The organizational surgeons, who are either hoping to cut or add new
organs to the lab are sharpening their knives with little knowledge of the
anatomy and health of their patient. Despite extensive data collection activity, little strategically relevant information is available. Why? In the first place,
much data-collecting activity is performed by private researchers with private
or academic objectives (such is the case with the NCRDP) that may or may
not come to the attention of policy makers, depending upon a number of
haphazard conditions. In the second place, much of the relevant government
information focuses on perceived problems as those problems come to a head.
This is especially the case with respect to the information developed by the
U.S. General Accounting Office. Even the science indicators effort at the
National Science Foundation (the sponsor for early NCRDP work reported
in Bozeman and Crow, 1988) is severely limited by its reliance on secondary
data and by being limited largely to in-house resources. What type of information *is* needed? We recommend that strategic information be compiled
that has the following characteristics:

Strategic Information Should be System-Based Rather than Sector-Based

Most questions about deploying federal laboratories cannot be addressed
effectively in the absence of information about nongovernment labs. In
almost every case, the answer to questions about the appropriate direction of
federal laboratories depends upon whether other performers are available to
provide the required technical goods and services. This is the point made
again and again in the part of this monograph concerning "new ways of

thinking about R&D laboratories" and in a variety of NCRDP publications pertaining to the Environmental Context Taxonomy (Bozeman and Crow, 1990).

We are not arguing here that the Environmental Context Taxonomy or some close variant be used for strategic policy decisions. That typology, while providing more detailed information than the sector approach, is insufficient. Much more detail is needed. Starting with one or more classification schemes based on the type of output and the technical activities of the laboratories, information needs to be provided about the particular scientific and technical foci, programs, resources, major equipment, and outputs of the laboratories. This information should be at a sufficient level of detail to permit the cataloging of technical activities of labs by technical topics (such as, for example, low temperature superconductivity, flat panel displays, or genetically altered food products). The information should be comprehensive, allowing identification of all the major performers for a broad set of technical activities.

PANEL DATA SHOULD BE PROVIDED FOR VIRTUALLY ALL THE MAJOR CONTRIBUTORS TO THE TECHNICAL ENTERPRISE

The system-level studies from the NCRDP have provided evidence that a vast majority of the approximately 16,000 U.S. R&D laboratories are small, engineering-oriented operations focused more on testing, improvement of manufacturing process, and incremental change than on innovation and technical changes affecting the NIS. Arguably, there are no more than 1,000 laboratories of sufficient size and resources to make significant contributions, on an ongoing basis, to public domain science and technology or to provide technological innovations. If the geological survey can provide information about thousands of minerals and the Census Bureau can provide vast information about the movement, habits, and activities of 270 million people, it should be possible to provide valid panel data, on a regular basis, for the 1,000 top U.S. R&D performers. With this systemwide information it should be possible to make more informed decisions about both the R&D investments and the optimal uses of federal laboratories. Indeed, such an inventory could be invaluable for policy making in a variety of science and technology-related realms. If such an inventory is developed, it may be desirable to dispense with other studies on a smaller scale.

Strategic Realignment of Laboratories' Technical Portfolios

The potential for *tactical* or operational realignment of federal laboratories and their technical portfolios is high (and to some extent is occurring at a rapid pace in the 1990s). The potential for *strategic* realignment is low. In this section we introduce three models of realignment, the last embodying our preferred approach. We also provide more specific recommendations about approaches to realignment.

We can distinguish among three models of realignment:

1. Decentralized Mission Expansion and Diversification (Decentralized Model)
2. Coordinated Mission Realignment based on Industrial Needs (Coordinated Model)
3. Balanced National Needs-Driven Realignment (Balanced Model)

Each of these models corresponds to a general approach to change. Each has advantages and disadvantages, but not all three are equally desirable approaches.

Decentralized Mission Expansion and Diversification

There is no question as to whether R&D laboratories *will* change, the question rather is *how* they will change. As the various taxonomic studies of the NCRDP make clear, environmental forces are constantly changing. Forces such as changes in macro- and microeconomic conditions and broadscale changes in types and degrees of political constraints and opportunities inexorably affect the system of U.S. R&D laboratories. In most evolutionary phases, change is not particularly self-conscious or self-evident. Much of the history of change in R&D institutions is one of institutional and organizational drift, only occasionally punctuated by a period of rapid and highly visible designed change, usually due to some "shock" to the system, such as war, extreme economic upsurge or downturn, or major changes in political regime. Even in periods of shock, much of the R&D system is resistant to change, in part because insufficient resources exist to quickly adapt.

Currently, the federal laboratory system best fits the decentralized mission expansion and diversification model of change. Under this model, change is rapid but largely out of control (at least from a system perspective) as units of the system make choices that are defensive in nature and responsive (though not necessarily in a rational way) to system shocks. The model is "decentralized" in the sense that the responses are from individual labs, whereas the shocks to which they are responding are environmental ones outside the lab's control. When missions are threatened, the primary response is to look for new ones without a great deal of thought given to the long-range strategic fit of the choices.

A good example of this problem is the struggle between Los Alamos National Laboratory and Lawrence Livermore National Laboratory for role definition in the post–Cold War nuclear weapons research arena. As research moves toward containment of nuclear materials, cleanup of the nuclear materials mess, nuclear weapons testing on computers, and testing of theoretical weapons, the almost 50-year contest between the two labs continues. Historically, these labs have battled for resources among the various U.S. groups wanting nuclear weapons for a variety of applications (Francis, 1996). The labs, therefore, evolved in a complex political/technical setting where the competing interests of the various military services and the Atomic Energy Commission/ERDA/DOE complex fought out the issue of lab direction. Since the elimination of most nuclear weapon testing in the 1990s and since our bilateral efforts with Russia to eliminate nuclear weapons have taken off, these weapons labs have continued their contest for dominance, with each lab trying to win a new mission role and major new facilities with the hope that jobs can be saved and lab strength maintained. These labs are not moving forward according to an assessment of their strengths and weaknesses or based on a plan for building national strength. They are moving forward in political competition with each other. There is no meaningful plan allocating responsibilities.

As laboratories independently expand into similar missions and into missions that do not fit their historic institutional niche, the result is confusion and inefficient use of resources in the NIS. Currently, "mission creep" is rampant in the federal laboratory system, especially among the largest laboratories. While many of the new activities are worthwhile, they may not be the most efficient or system-rational use of resources.

Coordinated Mission Realignment Based on Industrial Needs

To a large extent, most recommendations about realigning the federal laboratory system have focused on the potential of the labs to contribute to industrial competitiveness. According to this model, the realignment would be coordinated, but chiefly along one dimension and certainly not the dominant one in terms of current activities of the labs.

If one were to assess the laboratories' potential to actually contribute to industrial competitiveness and then act on the basis of that assessment, this model would represent some improvement over the current disarray and "mission creep" of the decentralized model. However, such an approach still would not address the opportunity cost elements of industrially oriented realignment. Thus, if a given federal laboratory is quite good at commercial work but the expansion of commercial activity undermines a unique scientific mission that is in the national interest, the realignment still is not "system-rational."

Balanced National Needs-Driven Realignment

A balanced model undertakes realignment with strong consideration not only of the possible commercial value of the federal laboratory system but also of the significance of the current scientific and technical activities of the laboratories. It is decidedly not the case that a hands-off attitude should be taken with the less commercially oriented scientific missions. In some cases, labs' scientific activities are neither of high quality, unique, synergistic, or facilitative. A realignment focused solely on commercial activity may well neglect to fully consider needs for less commercially oriented scientific and technical activities. Moreover, a balanced model would also consider scientific, technical, and commercial gaps. What national technical needs are not currently being met (or being met inadequately) in the NIS and how might they be met with a thoroughgoing realignment?

A balanced model differs from other approaches in that it is not reactive but proactive. That is, the questions and processes driving the realignment are those pertaining to system-level needs and the extent to which those needs are being met. In the decentralized model the chief concern is with the individual lab's needs.

Critical Factors Pertaining to the Redesign of Federal Labs

In previous chapters we provided some indication of the types of information needed to allow strategic, systems-oriented planning, policy making, and policy design for the federal laboratory system. In addition to those needs, however, there are other, more particularistic needs associated with each of the "critical factors" for realignment. We discuss information needs below in connection with each of these factors. *We recommend that each of the factors given below be viewed as critical to realignment and each be part of assumptions and approaches to realignment and the decision making preceding it:*

- Coordinating mechanisms that go beyond the current departmental organization and jurisdiction
- Greater autonomy: reduce the control of topical program monitors in the mission agencies
- Mission rationalization
- An "opportunity cost" approach to assessing the industrial assistance mission
- Creation of a "Federal Laboratory Realignment Commission" authorized to recommend labs to be closed, but also to decide on radical changes in existing laboratories and needs for new federal laboratories
- Job retraining and relocation programs for personnel not needed in the new, reconfigured federal laboratory system

Each of these critical factors is discussed below and more specific recommendations are offered in connection with each. Generally, the focus is on the recommendation and the reasons for the recommendation, not on its implementation. We recognize that it would be especially difficult (though not impossible) to implement some of these recommendations, but we assume that the stakes are high enough to merit full consideration of political feasibility.

COORDINATING MECHANISMS THAT GO BEYOND THE CURRENT DEPARTMENTAL ORGANIZATION

It will not be possible to change the federal laboratory system fundamentally unless one begins with the assumption that the current structure of departmental control of the laboratories will change. This is not to say that current departmental and agency affiliation are invariably dysfunctional. In many cases, the current jurisdictional arrangements make good sense. But in other cases the relationship between laboratories and agency or department parents is only a historical artifact and more a barrier to than facilitator of productivity. New organizational and interorganizational designs need to be arranged. We recommend the following:

1. Generally, considerations of organizational affiliation of laboratories should be based on the following questions: (a) Is the preponderance of technical activity of the laboratory in direct service to the agency's mission? (b) Does the technical activity serve a clientele traditionally associated with the agency and its mission? (c) What specific "value-added," if any, does the department or agency bring to the technical mission of the laboratory?

2. Mechanisms must be created to enhance the laboratory-to-laboratory linkages (which often are weak) rather than the laboratory-to-agency program monitor linkages (which often are too strong and create counterproductive dependencies).

MOST FEDERAL LABORATORIES NEED TO BE GIVEN GREATER AUTONOMY

Virtually all the effectiveness indicators from NCRDP show that laboratories are more effective when they are more autonomous. Interestingly, this finding pertains to both the fundamental science mission (Crow and Bozeman, 1987a) and commercial missions (Bozeman and Coker, 1992). Thus, we recommend:

1. Increased laboratory control in technical agenda-setting, with budgets provided within a very broad programmatic framework

2. Continuing the trend of allowing laboratory directors to make decisions about ties to industry and universities through CRADAs and other cooperative agreement vehicles

3. As above, decreased influence by topical program monitors

4. But also increased accountability through thoroughgoing biannual external evaluation

Missions Need to be Rationalized

Owing to a number of factors, including "mission drift," mission imperialism (or entrepreneurialism), and unclear or unworkable statutory authorization, there are relatively few large federal laboratories with clear, focused missions. Ambiguity of mission and objective typically results in a variety of performance problems ranging from red tape (Bozeman and Crow, 1990) to inadequacy as a commercial partner (Bozeman, Papadakis, and Coker, 1995b). We recommend:

1. A zero-based assessment of mission needs should be undertaken separate from any consideration of existing capabilities and deficits. If one were starting from scratch, what missions should the federal laboratory system serve?
2. Missions need to be problem-directed rather than research directed
3. Some of the largest federal laboratories, particularly public science labs, should be given the status equivalent to independent agencies and be provided specific charters to perform pure basic and directed basic research. This would entail new statutory authority whereby the handful of national science centers would have their own lines in the federal budget. These laboratories should be commissioned only when it is possible to demonstrate: (a) world-class research excellence; (b) strong needs for expensive equipment and resources; (c) an ability to work closely with university and industry-based basic researchers; (d) a valid, university-style set of methods for evaluation of scientific quality.
4. A much larger set of laboratories should be designated as technology development laboratories and should be either placed under Department of Commerce jurisdiction, departmental or agency jurisdiction if the primary product is to be used by that department or agency, or should be contractor-managed, hybrid technology labs. These laboratories should be commissioned with the understanding that their chief role is to contribute to industrial competitiveness, and they should be evaluated by industry-based panels.

An "Opportunity Cost" Approach to Assessing the Industrial Assistance Mission

Most assessments of federal laboratories' technology transfer and commercialization missions have proceeded based on an implicit "out-the-door" model or a "market impact" model (Bozeman and Fellows, 1988). The "opportunity cost" model assumes that the success or failure of laboratories' technology transfer investments should be considered in terms of alternative investments and in terms of the impact of the mission on the laboratories' overall capabilities. NCRDP research has shown very different results depending on the assumptions and the evaluative model employed (Bozeman and Coker, 1992; Bozeman, 1994a,b). The evidence shows that "one size fits all" technology policies are misdirected. If a laboratory increases its technology transfer effectiveness by 10% but its ability to contribute to public domain science is reduced by 20%, this may not be a great accomplishment. We recommend:

1. Among multipurpose laboratories, assessments of capabilities and effectiveness should be focused on the entire array of technical activities and missions, not just on scientific accomplishment or commercial success.

2. Careful analysis must be undertaken to determine the effects of commercialization activities on basic and applied research (and vice versa). Currently, there is very little information on that topic and some evidence (Crow and Bozeman, 1987b) suggests that there are circumstances where effectiveness in one mission undermines effectiveness in another.

CREATE A "FEDERAL LABORATORY REALIGNMENT COMMISSION"

Among the many commissions and blue-ribbon panels convened in recent times to examine the status of the federal laboratory system, the Galvin Commission (U.S. DOE, 1994b) seems to have been set up with some expectation that its recommendations might actually be implemented. However, the Galvin Commission, despite its prominence and the attention it has been given, still had a relatively narrow mission. First, it examined only a fraction of DOE laboratories. Second, it eschewed being a "base-closing commission for federal laboratories."

What is needed is a carefully constructed federal laboratory commission that has a much broader charter and a willingness, if evidence supports such a decision, to close labs and to suggest the radical redeployment of others and even the opening of new labs. While there seems little likelihood that personnel within the federal laboratory system would ever be enthusiastic about a process that would almost surely result in the closure of some federal laboratories, it is quite likely that the scientific and technical community at large could support such a process *if* its members were convinced that the objective was to provide no net loss (and perhaps some gain) in the federal investment in science and technology. Thus we recommend:

1. A "Federal Laboratory Realignment Commission" should be created to oversee a zero-base assessment of federal laboratories with a view

to: (a) closing some laboratories; (b) recommending organizational and jurisdictional changes (of the sort discussed above); (c) recommending redeployment of laboratories (including downsizing when necessary to sharpen the missions); (d) recommending the filling of gaps in the NIS by creating new federal laboratories (or by channeling science and technology resources into other institutions that can serve specific scientific and technical needs not being met at present); (e) investigating possible new institutional organizational designs for achieving scientific and technical objectives in the national interest.

DEVELOP A JOB RETRAINING AND RELOCATION PROGRAM FOR EXCESS LABORATORY PERSONNEL

Even if one assumes that laboratory realignment will result in little or no net loss in the number of scientific and technical personnel employed by government (which is probably *not* a realistic assumption) there will still be a need for a considerable job retraining and relocation program for federal lab employees. Even if new labs are created there is no reason to believe that technical needs will correspond completely to the skills and capabilities of those displaced by the closing or reconfiguring of existing federal labs. In most cases, the assumption that today's bomb-maker is tomorrow's environmental engineer is, at best, wishful thinking.

There are two good reasons to initiate a job retraining and relocation program. In the first place, if such a program is not developed, the likelihood of making changes in the federal laboratory system are greatly diminished. In the second place, and more important, it is the right thing to do. The fact that hundreds, perhaps thousands, of laboratory employees are not now needed to perform tasks that were once needed, even vital, is certainly no excuse to discard redundant people as we discard redundant missions. We recommend:

1. Funds saved from closing or cutting back federal laboratories should be earmarked for job retraining and relocation programs for displaced employees.

2. After retraining and relocation programs have been fully implemented, savings from federal laboratory closures and downsizing should be designated for new federal (or hybrid) laboratories focusing on missions identified by the commission as being the highest priorities for serving national scientific and technological needs.

Industry's Perceptions of Federal Laboratories as Commercial Partners

The most recent NCRDP industry-based study (Bozeman, Papadakis, and Coker, 1995b) provides only a limited perspective on industries' views of federal laboratories because it is limited to those companies that have interactions with the federal laboratories. The issue, however, is what is the perception of U.S. industry as a whole?

Government is rarely viewed as a major source of technical information. Even today, after several years of effort designed to attract industry as a collaborative technology partner, industrial technology developers still do not look to federal labs as a significant source of technical knowledge. Research by Roessner and Bean (1991), in a study focusing on the larger companies that are members of the Industrial Research Institute, showed that federal laboratories are well behind vendors, suppliers, universities, and industries' own R&D operations as a source of technical knowledge. However, the recognition of federal laboratories as a source of technical knowledge has, apparently, increased somewhat in the past few years.

The perceptions of companies actually working with federal laboratories are favorable. Research results (Bozeman, Papadakis, and Coker, 1995b) indicate that the customers of federal labs are, for the most part, quite satisfied. According to a global measure of satisfaction with the experience, fully 89% agree or strongly agree that the commercial interaction was a good investment of company time and resources and that the company would work with the federal lab again given an appropriate opportunity.

Partly as a result of these findings, we recommend:

1. Rather than focusing substantial resources on "recruiting" industrial partners, federal laboratories should focus their efforts on companies that have a clear predisposition to work with them. While it is useful to attract particular companies to particular technological opportunities, it is generally not an efficient use of resources to "convert" companies that have little general inclination to work with federal labs; the ideological and cultural barriers are likely to prove too strong.

2. More attention should be given, by both federal labs and policy makers, to the extent to which the federal laboratories' technical knowledge complements or substitutes for other sources of technical knowledge. If the result is that technical research or company technical investment is reduced, then laboratory-company interactions may be net positive.

An issue that has received much less attention than industry's interest and experience in working with federal labs, is the companion question, "what is the effect upon industrial laboratories of increasing government involvement?" Perhaps one reason this question has received little attention is that the number and range of government effects are so pervasive as to be taken for granted. Thus, while collaboration with federal laboratories has received much attention, the fact is that only a small fraction of industrial researchers have any contact whatsoever with federal laboratories. But virtually every large industrial R&D lab is affected by changes in the tax code (including but not limited to those directly aimed at R&D), government regulations, and allocation of government resources through research contracts and government purchasing. Evidence points to the fact that government involvement has considerable and even predictable effects on industry (Bozeman, 1987; Perry and Rainey, 1988; Bozeman and Bretschneider, 1994).

Measuring Federal Laboratories' Commercial Performance

The question of how to measure federal laboratories' commercial effectiveness requires answers to several antecedent questions.

ONE LAB OR MANY, ONE PROJECT OR MANY?

Generally speaking, techniques that are useful in the aggregate may be less useful in the individual case (and vice versa). This is a simple point but an important one. For example, a case study approach is likely to provide great insight into the effectiveness of particular projects, especially if the rigors of case method (Kingsley, 1993) are met. While case studies can be aggregated such that multiple projects and organizations can be evaluated with case methods (Kingsley, 1993; Kingsley, Bozeman, and Coker, 1995b), to do so is quite expensive and time-consuming and, even with cases undertaken with common research protocol, inevitably challenges the researcher/evaluator to provide a valid approach to cross-case aggregation. But in small sets of projects or single projects, case approaches can have a high degree of validity. Typically in case study approaches "more is known, but about less."

While one would not wish to forego the level of detail and contextual information provided by the case study analysis of individual (or small groups of) projects, that level of detail can be stifling and even wasteful in large-scale studies. Moreover, in aggregate analysis it is possible to exploit a number of the attributes of probability theory in sampling and inferential statistics such that "less is known, but about more."

MEASUREMENT FOR WHAT?

The best approach to measuring effectiveness depends, to a considerable degree, upon the intended use of the measures. Among these uses are: program monitoring, planning, resource allocation decisions, program improvement, documentation, and marketing. Various purposes require different levels of precision, detail, and accuracy and different degrees of confidence in the validity of casual assumptions. Roughly speaking, we can differentiate between "high confidence" and "low confidence" approaches, recognizing that these are two poles. "Low confidence" approaches are useful under the following conditions:

- Results are needed very quickly
- There are limited resources to devote to the evaluation
- There is no reason to impute causality
- An imprecise estimate is sufficient

There are many circumstances that meet the requirements for "low confidence" approaches. Sometimes only a rough answer is needed to the question "does this work?" while there is no immediate need to know the exact extent to which a program works or exactly why it works. Often low confidence approaches are useful for midterm adjustments in programs. On the other hand, high confidence approaches are appropriate in the following cases:

- The evaluation will support significant decisions
- Determining causality is important
- Monetary resources are available as well as human resources (i.e., technically trained evaluators)
- There is ample time to execute a carefully designed research plan.

In most instances, there is no need for the extreme variety of high confidence research. Moreover, even when there is a need, there may not be the time and resources to support such a study. Thus, high confidence approaches should be used sparingly and should not be used at all if there is any question that the resources and time needed will not be available. Most important, low confidence approaches should never be cloaked in high confidence trappings—it is vital that the requisite precision, accuracy, and validity match the evaluative problem. In all too many cases, statistical or mathematical sophistication is inappropriately taken as an indicator of "high confidence" evaluation research design when it is merely an element, and rarely among the more important ones, of reliable evaluations.

MEASUREMENT FOR WHOM?

The selection of measures depends, at least to some extent, upon the characteristics of the audience. Given the extent to which measurements depend upon the characteristics of the audience and given the extent to which measurement is linked to method and theory, audiences that are more interested in "a number" than in an explanation will require different approaches than those interested in both a degree of effectiveness and the causal mechanics of effectiveness.

The above considerations provide a context for understanding why the NCRDP researchers have been concerned as much with "why" issues as with "how much" issues, why they have focused on aggregate levels (all labs or large groups of labs), and why they have targeted a mixed audience including both researchers and policy makers. Thus, for the purposes of NCRDP research, relatively "high confidence" techniques were in order. These techniques would permit some degree of causal inference and explanation. The problem, however, was that aggregate studies, especially questionnaire-based studies, would often experience threats to construct validity. There is a limit to what can be measured in mailed or phone-administered questionnaires.

As with all methodological approaches, survey research has clear-cut advantages and disadvantages. Recognizing the need to consider the specific advantages and disadvantages of a variety of techniques employed to evaluate R&D impacts in general, and technology transfer in particular, NCRDP researchers (Bozeman and Melkers, 1993) published a compendium overview of R&D evaluation techniques. While there is no need to recapitulate the basic premises of a book available in the literature, it is instructive to consider the measurement and evaluation techniques reviewed in that book in terms of their advantages and disadvantages for the more specific question of technology transfer and commercialization effectiveness.

The study considered the more common approaches to R&D impact evaluation in terms of the following criteria: cost, time required, level of internal validity, external validity, statistical and measurement precision, accessibility to nontechnician audiences, and specific relevance to questions of technology transfer and commercialization effectiveness. It is useful to consider the question of the relevance of these approaches to questions of technology transfer. In our view, the approaches most useful for technology transfer are (1) survey research and questionnaires, (2) econometric and internal rate of return approaches, and (3) case studies. The utility of survey research and questionnaires is derived from the fact that they permit convenient and reasonably valid data gathering from large aggregations. Econometric approaches tend to be extremely useful in those cases where benefits are largely internalized (i.e., specific firms' benefits are measured) and less useful when there is a concern with externalized and public domain benefits (which are extremely difficult to measure using such techniques). Case studies are potentially the most useful approach, especially when measuring benefits accruing from particular projects. But case studies tend to be quite limited beyond small aggregations of projects.

Approaches that have an intermediate level of utility are (1) bibliographic approaches, such as citation and patent analysis, and (2) most approaches to peer review. In most cases, commercial activity in federal laboratory-industry interactions does not revolve around patenting. While citations may be a useful (though limited) surrogate indicator of scientific impact, they are of very little, if any, value for assessments of commercial impact.

Thus, bibliographic analysis is limited to applications focusing on cross-impacts between scientific output and commercial effectiveness and on those commercial interactions revolving around patenting and licensing activity. Conventional peer review is limited in its applications for questions of technology transfer, but in a pilot study Kingsley and Bozeman (1992) determined that peer review panels could be employed with some success to use expert opinion to assess the effectiveness of technology transfer projects.

An extremely important issue in assessing the effectiveness of federal laboratories' technology transfer and commercial activities concerns the concept of effectiveness. NCRDP researchers have used a typology (Bozeman and Fellows, 1988) distinguishing among the following models: "out-the-door," "market impact," "political," and "opportunity cost" models. Empirical research (Bozeman and Coker, 1992; Bozeman, 1994a,b) has shown that the four different effectiveness models require different measures, are only moderately correlated with one another, and are explained in terms of different sets of determinants. While these are not the only possible conceptualizations of technology transfer effectiveness, the lesson is clear: "effectiveness" is multidimensional and, in some cases, one type of effectiveness is achieved at the expense of another. In the absence of clear conceptualization, effectiveness measures contribute little.

We have a number of recommendations regarding approaches to measuring the effectiveness of technology transfer and commercialization activities of federal laboratories.

1. Be explicit in the choice of effectiveness models and criteria, "effectiveness" is multidimensional (Bozeman and Fellows, 1988). In most cases, the "opportunity cost" model is particularly appropriate in that it considers not only what is achieved by commercial activity but also the context of possible alternative uses of the resources employed to achieve commercial goals.

2. Other things being equal, multiple method approaches to evaluation are to be preferred. As is clear from the above discussion, different R&D impact methodologies offer different sets of advantages and disadvantages. Using multiple methods generally increases the likelihood of achieving valid results.

3. Relatively low-cost, "low confidence" technology transfer evaluations should be routinely undertaken at federal laboratories, using commensurate measures. Generally, these should focus on individual projects and programs at individual labs. The Department of Commerce should commission periodic "high confidence" studies of technology transfer that are performed by researchers not affiliated with federal laboratories but involve both lab personnel and actual and potential

industrial partners. These studies should be conducted for sets of laboratories and multiple projects and programs. Before conducting the effectiveness research, an implementation plan should be in place to assure that these relatively high-cost studies are used in decision and allocation processes.

Federal Laboratories' Commercial Effectiveness: Results and Recommendations

Chapters 5 and 6 provided information about the current level of effectiveness of federal laboratories' commercial activities. Those findings suggest that although there is great variance in the commercial effectiveness of federal laboratories, generally, the laboratories have an important role to play in commercialization of technology and that some seem to be playing that role quite well. But many other laboratories have virtually no commercial activity. It is important to remember that the NCRDP study of industry's evaluation of the benefits of federal laboratory interactions was limited to "commercially active" laboratories and that, even by a lax definition of that category, few were so categorized. Furthermore, Bozeman and Coker (1992) found that in a larger and more diverse sample of federal laboratories, only a small fraction was commercially active. Most had no patents and no licenses whatsoever, and the median number was less than two. Thus, an issue prior to the level of effectiveness is simply the amount of activity, and most data indicate that the level of commercial activity has been, at least until recently, limited. However, focusing on the commercially active federal laboratories, have they been effective? NCRDP findings can be summarized as follows:

- Of the interactions between federal labs and industry examined by Bozeman, Papadakis, and Coker (1995b), about 20% had already led to a product being marketed. This despite the fact that many of the projects were ongoing and in 79 cases products were still under development. In 52 cases existing products had been improved.
- Federal laboratories fare poorly as job creators, at least from the standpoint of jobs created as a result of industrial interactions (Bozeman, Papadakis, and Coker, 1995b). On average, 1.6 new jobs were created by interactions between federal laboratories and industry.
- The average level of benefit reported was $1,604,564. The average level of cost reported was $417,010. At first glance this looks as though the companies had a 4:1 multiplier for their interactions with the federal laboratories, but it must be kept in mind that in many instances the companies have essentially no cost for the interaction.

- However, this figure masks a good deal of diversity. A significant percentage (26.6%) of the companies did not value the benefits as highly as the cost or, in effect, did not break even in their investment. About 10% of the respondents report no benefits from the interaction.

As a result of specific findings regarding the level of effectiveness of federal laboratories, their technology transfer, and other commercial activities, we recommend the following:

1. It is imperative to recognize that only a fraction of the federal laboratories have real potential to contribute effectively to technology commercialization goals. Steps must be taken to identify those that are effective and to avoid burdening other laboratories with commercial activities for which they have little hope of effectiveness. In particular, federal laboratories with unique or scarce equipment, resources, and know-how should be among those chartered to engage in technology transfer and commercialization activity.

2. There is no reason to expect that federal laboratories will be able to play a significant role as job creators, and thus assessments of commercial effectiveness should focus on other economic measures.

3. Technology transfer policies should not focus on companies in an undifferentiated way. Rather, they should take into account the attributes of companies likely to be successful technology-development partners. Among these, the R&D intensiveness of the company should be given especial attention.

What Federal Laboratory Technology Transfer Strategies Seem Effective?

NCRDP research has explicitly addressed the question of which technology transfer and commercial activities appear to be the most effective (Bozeman, Papadakis, and Coker, 1995b; Bozeman, 1994a,b). Federal laboratories have used a number of strategies in their technology transfer and commercialization efforts, including licensing, resource sharing, consortia, dissemination of literature, conducting workshops, and personnel exchange, just to name a few. Until recently, there was very little evidence regarding which of these strategies were effective or even the extent to which each is used, but that void has been filled to some degree. Based on NCRDP results, we recommend:

1. Federal laboratories should begin to "market" their most valuable product, directed basic and applied research, and tone down claims about the value of existing technology.

2. The concept of technology transfer should be reconsidered since it connotes a relatively passive, noninteractive role between companies and federal laboratories; the terms "partnership" and "cooperative R&D" are more realistic and more useful signals regarding what to expect from the relationship.

3. Those federal laboratories that score highest on industrial partners' value ratings should be considered for designation as industrial competitiveness laboratories and should have substantial increments in their commercial activities. This money should be a transfer from laboratories who are by all indicators clearly ineffective in their technology transfer and commercialization activities.

4. A national fund should be established to be distributed to laboratories that have highly rated technology transfer and commercial activities. Regardless of agency or departmental affiliation, these awards should be made by a panel that includes both laboratory personnel, industrial partners, and independent evaluators.

How Should Federal Laboratories Contribute to Competitiveness and Other National R&D Needs?

One "answer" is derived from the way we have formulated the question, adding the clause "and other national R&D needs." By this point, it is clear that our approach is based on the opportunity cost model. It is also fundamental to recognize the relation between commercial and "competitiveness" objectives of policy makers and labs to the many noncommercial missions already served, often quite well, by the federal laboratory system.

1. Rationalize labs so that they have clearly defined roles. Some may focus on commercial support, some on basic research. But they cannot be all things to all people.

2. Develop ways to clearly communicate missions to individual laboratories. The communication of findings is less important than the communication of capacity and specialization. The best way is to keep it simple so that the message does not get muddled. Currently, industry has only sketchy knowledge about the particular strengths of particular labs.

3. Labs designated as commercial should be radically transformed in structure; they should have governing boards that include industry representatives. Such labs should have highly active personnel exchanges. Legislation should be created to provide an industrial analog to the Intergovernmental Personnel Act, facilitating personnel exchange between commercially oriented federal labs and industry.

4. The Japan Research and Development Corporation (JRDC) provides grants, regardless of sector, to facilitate the development of technology that is licensed by their government research institutes (Bozeman and Crow, 1991a). These grants are highly flexible and aimed specifically at bridging the gap between the applied research stage and the birthing of a product or processes. A similar system of sector-blind grants can greatly facilitate commercialization of federal laboratory technology and (in the process) enhance the value of federal lab licenses.

5. Industry development labs need to be built in twos so as to compete with one another. Then those not successful by performance criteria should have their resources redeployed. Interestingly, this principle of competition was used in the weapons labs, particularly pitting Sandia and Lawrence Livermore against each other, and the same controlled competition can be used to stimulate commercial activity.

6. Even in today's markets it is not always easy to attract the best technical and entrepreneurial talent to the federal laboratory systems. Policies should be developed that provide for short-term, fixed-year (but relatively lucrative personnel exempt) contracts to attract the best young scientists and engineers to commercially oriented government labs and expose the technical staff to industrial partnerships. The explicit objective being that these "short-termers" will migrate to industry after the lapse of their contracts.

Summing Up: Toward a Less Mysterious 16,000

When we began our R&D laboratory studies nearly fifteen years ago, we assumed initially that our labors would contribute to a vast literature of empirically based studies. Granted, we knew of only one study of this sort, Frank Andrews's (1979) UNESCO-sponsored, cross-national study of R&D units. Nevertheless, we were sure we would find many more once we began in earnest to dig through the archives. We found instead a great many studies focused on one or another aspect of R&D laboratories, including a profusion of studies of DOE multiprogram labs, case studies and histories of particular labs, government-sponsored compendia of data about scientists and engineers, and a few publications decrying the lack of data about R&D labs.

Once we had discovered the lacuna, we moved slowly in our efforts to close it. Our first efforts aimed at enhancing organization theory (e.g., Bozeman, 1987) more than at contributing to public policy. But as we gathered more data about R&D laboratories we began to accumulate some insights not widely available. Once we found how easy it was to dazzle our colleagues with such arcane information as the approximate number of R&D laboratories, we were off and running. We smiled to ourselves as we read articles in the *New York Times* on licensing activities at federal laboratories ("we actually have some idea of how many laboratories hold licenses!") and grimaced as we read articles berating laboratories we knew to be among the most productive in the world. Our objective crystallized: we would be the Linnaeus of R&D laboratories. After managing to convince a few government agencies that knowing more about R&D laboratories would be a good thing, we plied our trade. Having now laid out a good part of that trade in the previous pages of this book, we return to the core issues and objectives, including:

- Are the mysterious 16,000 now less mysterious? What next? What should we (and others) take as next tasks in policy analysis for R&D laboratories? What gaps remain in our data and theory and in its application?
- What is the value of the institutional design approach? How can the approach be extended?
- What is the future of the U.S. R&D laboratory system?

DEMYSTIFYING THE 16,000: WHAT NEXT?

Yes, the NCRDP has succeeded to some degree in providing information relevant to the 16,000 U.S. R&D laboratories. Despite some danger of developing (from five separate projects) so much data that digesting it is impossible, the NCRDP provides a picture of R&D laboratories well beyond the state of our empirical knowledge in 1983. As a result of NCRDP studies and others' work, we have greater knowledge of the range of activities of R&D laboratories, their institutional and environmental positioning, cross-national comparisons (not dealt with here but in several published reports and articles), and even factors pertaining to certain types of effectiveness, especially technology transfer and commercialization.

Nevertheless, the 16,000 continue to hold many secrets. The NCRDP has shed some light on the U.S. R&D laboratory complex as a system, but many significant aspects remain mysterious. We do not have nearly enough information about interrelationships among laboratories and among laboratories and the organizations they serve. As R&D in the U.S. increasingly becomes a multiorganizational enterprise, knowing about aggregations of labs without knowing about their interactions will take us only so far. Thus, the 16,000 parts of the U.S. R&D laboratory system are more visible than before the NCRDP, but the workings among those parts, the dynamics of the system, still remain all too mysterious. More work is needed on networks of laboratories and their interorganizational relationships. This work has already begun in other countries, albeit on a subsystem scale (Niosi and Bergeron, 1992; Callon et al., 1992; Joly and Mangematin, 1996).

Despite our efforts to understand the dynamics of change in the U.S. R&D laboratory system, we have managed to shed little light on that topic. In moving from Phase II to Phase III of the NCRDP, and charting many of the same laboratories, we hoped to develop knowledge of system change. Unfortunately, the time span examined was too compressed (just four years), our instruments (mailed questionnaires) were too blunt, and our panel data too problematical. The chief finding, not unimportant, pertained to the rate of R&D laboratory "mortality." But an understanding of system change remains vitally important and cross-sectional data contributes little to that understanding. If one subscribes to the *never neutral principle*, that changes in public policy translate to far-reaching shifts in focus and composition of R&D laboratories, the need for time-series, panel data seems apparent.

The task of monitoring 16,000 R&D laboratories seems unlikely to soon rise to the top of any government agency's agenda. If one throws into the mix the need to develop time-series data and information about interorganizational relationships, then only the most optimistic among us would have high expectations of any single institution undertaking the challenge. But two factors give hope. First, the power of the central limit theorem shows (Congressional opponents of census sampling notwithstanding) that a quality representative sample serves quite well in representing R&D laboratories. Second, the *player principle* argues the need to focus on a relatively small subset of the 16,000. In all likelihood, at least 10,000 U.S. R&D laboratories have little impact on innovation. Careful monitoring of the players requires relatively modest resources and, we argue, would yield much better thought-out public (and private) policies.

THE INSTITUTIONAL DESIGN POLICY MODEL: WHAT NEXT?

Less a policy-making paradigm than a model for policy analysis, the institutional design model provides two benefits. One is a rough analytical approach differing from the standard microeconomic analysis assumptions of the market failure paradigm. While certainly not competing with the market failure paradigm on the level of development or generality, the institutional design model at least compels attention to the system, not just its parts (i.e., the *systemic principle*). Institutional design further has the advantage of simplicity (at least in concept); it is so simple, in fact that it can be summarized with aphorisms. Let us take a forester metaphor to demonstrate the simple but nonetheless practical injunctions from the institutional design approach. Those wishing a bountiful harvest from U.S. R&D laboratories might bear in mind the institutional design approach to forest management:

1. If lost in the vastness of the forest, get your bearings from the tallest laboratory trees: the *player principle.*

2. But be aware of the forest ecology and the forces governing it: the *systemic principle.*

3. As you go into the forest, know that all you bring with you can affect the trees and their ecology, sometimes in unexpected ways, usually for good *and* ill (forest fires and lab closings being excellent examples of creative destruction): the *never neutral principle.*

4. The cabbage palm makes a nice lunch, cedar a handy bug repellent, oak a sturdy chair (but if you ingest cedar, repel bugs with oak, and build your house from cabbage palms, you lose your forestry merit badge): the *comparative advantage principle.*

5. If you chop down a white birch to make a dugout canoe, do not expect the tree to provide shade in the summer: the *opportunity cost principle.*

Having now tested the limits of human endurance of aphorisms and metaphors, we suggest that further work on an institutional design model must focus on verifying and sharpening its assumptions. For example, NCRDP findings already give support to the *player principle* by showing that the vast majority of R&D laboratories do not have as an objective contributing to the NIS and, instead, provide limited, proprietary technical outputs. But even if we are confident that a minority of R&D laboratories contribute significantly to the NIS, it is still necessary to know much more about the nature of players' contributions, the institutional factors that position them as players, and the nature and rate of change in the contributions of players. To what extent can they be mobilized and how quickly? Can labs outside the NIS be brought in? Can new players be invented? How much capacity and what kinds can players lose and still remain prominent in the NIS? It seems to us that the *comparative advantage principle* particularly warrants greater attention and more detailed forms of analysis. We have said at every turn that U.S. R&D policy needs targeting, that current approaches wield an ax when a scalpel is needed. But targeting requires not only an assumption that different laboratories do different things well, it requires knowledge of particular things and of the features that make laboratories strong. Presently, some observers have knowledge of the particular strengths of several laboratories, but such knowledge is disaggregated and hit or miss.

In sum, the institutional design model is not a theory of policy making or even a set of interrelated assumptions yielding specific prescriptions. Instead, it is a broad interpretive framework providing points of reference and rough guidelines. But our NCRDP analyses suggest that those rough guidelines, when followed, have the prospect of yielding some insights not likely to come from traditional, more familiar approaches.

R&D Laboratories: What Next?

Finally, we don our pointed, star-emblazoned wizards' hat and ask "What is the future of U.S. R&D laboratories?" We are particularly concerned with government R&D laboratories simply because those are the ones that policy makers can quickly and directly change.

The four science and technology policy paradigms frame much of our analysis, and perhaps fittingly, we conclude with projections based on those paradigms. Specifically, what might one expect during the next twenty years or so should one paradigm be fully embraced *to the exclusion of the others*? Obviously this is a highly unrealistic question—there being almost no chance that any of the three current policy paradigms will disappear as a major set of assumptions influencing policy. But even if unrealistic, the exercise seems a good means of furthering debate. By considering possible impacts of an exclusive paradigm, we can perhaps anticipate results of shifts in the relative strength of the respective paradigms. We focus chiefly on federal laboratories but give some attention to other R&D laboratories.

Scenario: Market Paradigm Dominance

The market paradigm suggests a modest role for federal laboratories—work within the context of missions not easily privatized. Depending on the stringency of one's view as to what is and is not easily privatized, the market paradigm ranges from a focus on that most public of public goods—national defense—to a variety of quasi-market public goods with strong externalities, such as public health. Moreover, the market paradigm holds little truck with energy or agricultural missions for laboratories and is immune to arguments that energy and agriculture missions not performed in the private sector need performing in the public one. There is not yet much evidence that today's proponents of the market paradigm have strongly embraced environmental research; it clearly has substantial externalities, and there is no evidence that the market will or even could produce an optimal level of environmental research. If there is a shortfall of environmental research today, that is probably due to the high threshold of indifference of many policy makers and R&D sponsors rather than to the ideological premises of environmental research. Unless pollution and hazardous waste problems are checked, or unless pollution credits and other such market solutions begin to have more of an effect than heretofore, the growth of environmental research could occur even within the framework of a market policy paradigm.

In, say, 2020, a federal laboratory system dominated by market paradigm thinking might have the following characteristics: (1) a smaller but stable set of federal defense-supply labs (including the phasing out of the weapons labs' "peripheral" missions); (2) the end of the federal energy R&D laboratory focus; (3) some support for laboratories focusing on quasi-public goods such as environment and hazardous waste and public-health-related R&D (but not medical research and technology, in general). Laboratories oriented to testing and standards-setting (including what is now the NIST system of laboratories) would be phased out under the assumption that the market, by its own devices, can produce standards and dominant designs. Cooperative industrial research at federal laboratories and laboratories' technology transfer activities—"government interference in the market"—would disappear. In sum, the role of the federal laboratory system would shrink dramatically and the interesting questions would pertain to the reallocation of billions of dollars or R&D funds (redirected to other providers, such as universities? To non-R&D missions? To tax cuts or tax credits?).

Even were the market failure paradigm embraced fully, the impact on industrial laboratories would, in all likelihood, be mixed. While federal labs' exit from such fields as energy research and technology could conceivably provide new market opportunities for industry, there is little reason to believe that the current lack of industry activity is due to a "crowding out" by federal labs. Further, the federal laboratories' current activities often fuel industrial labs. By taking research problems at the high-risk, high-cost phase, federal laboratories often make a net positive contribution to industrial research and productivity.

SCENARIO: MISSION PARADIGM DOMINANCE

In some respects, forecasting the effects of mission paradigm dominance presents the most problems. If we assume that federal agencies should be servants of line agency functions, we must know about the configuration of those functions—that is, we must read political tea leaves. If we hazard a straight-line trend extrapolation, most mission functions will continue to contract. At least as measured by numbers of government employees, the size of government has shrunk continually during the past ten years. Likewise, the growth *rate* in the federal budget, in terms of the number of federal programs, peaked more than two decades ago. Until very recently, the chief trend in federal missions was the growth of the transfer payments segment of the budget and the relative shrinkage of almost every other major category, excepting interest on the debt. But as the rate of deficit growth declines, inroads are being made into the transfer payment (read "welfare reform") aspects of the budget, and it appears that ours is a period of change in the strength of federal government missions.

There is nothing in current political trends to suggest an impending expansion of the three missions that have historically underpinned the largest portion of federal laboratory R&D—defense, energy, and agriculture. Thus, perhaps the "easiest" forecast is that reliance on a mission paradigm will, as with the market paradigm, result in a diminished federal laboratory system, accompanying diminished core missions. The likely exception is the defense and national security mission, which requires enormous expenditures just to preserve the base investment. Further, the microscopic "peace dividend" and the sound and fury of the base-closing commission provide ample testimony of the resilience of the defense R&D establishment.

Both universities and industry benefit from the mission paradigm focus, especially as the size and scope of federal laboratories continues to diminish. Without an upward trend—which is highly unlikely—in federal laboratory support, mission R&D will continue to be contracted out to industry and universities, and each laboratory type will continue to benefit from the mission paradigm and its consequences.

SCENARIO: COOPERATIVE TECHNOLOGY PARADIGM DOMINANCE

It is not easy to imagine scenarios by which the cooperative technology paradigm would become dominant, but there is perhaps one path—a more dramatic economic decline than the one that initially provoked interest in cooperative technology and "competitiveness." If the U.S. economy began to experience sharp declines (well beyond those of the 1980s) vis-à-vis its trading partners and if the U.S. balance of technology-goods payments was to become highly unfavorable, then policy makers might, out of a sense of desperation, be drawn to cooperative technology as a dominant framework. The likelihood further increases if an even larger number of U.S. competitors (not just Japan) implemented a strong set of highly effective cooperative technology policies.

If the federal laboratory system came to be viewed, in reality as well as rhetoric, as a major source of innovation for U.S. competitiveness, the system would be radically different under cooperative technology policies. In the first place, the traditional mission-agency basis of laboratory management would almost certainly be discarded as outmoded and dysfunctional. Federal laboratories as innovation centers would require either more autonomy (perhaps as independent executive agencies) or reorganization under a technology policy cabinet level department. A tough-minded policy aimed at technology exploitation of the laboratories for national economic development would probably set aside the majority of activities now undertaken as CRADAs and cooperative research. So long as cooperative technology is an ornament to other missions or a political symbol, resources go to as many claimants as possible with limited concern about impacts. But a serious economic development effort would require triage—the federal laboratories would have to devote resources to companies, large and small, that demonstrate a capacity to exploit technology. Many of the mom-and-pop companies now working with the labs do so either for the mutual benefits of public relations or for the purpose of using lab assets for resource substitution and cost-cutting for the companies. With a serious cooperative technology policy in place, companies cooperating exclusively for these reasons would no longer be served. Likewise, the large corporations that sometimes devote resources to federal laboratory partnerships with no real monitoring or expropriation of results, would be cast aside.

Gauging the likely impact of cooperative technology development on industry and university labs poses problems. First, the impact would depend upon the readiness of nongovernment labs to cooperate. Current evidence suggests that industry labs approach cooperation with great caution. But providing incentives, including matching R&D funding, could well alter industry's receptivity.

But the cooperative technology paradigm will remain a hard sell, in part because of the coolness of U.S. industry to government, but also because there are few undisputed big successes coming from the policies and projects of the paradigm. One explanation of why the Manhattan Project analogy has so often failed with respect to other technology development efforts is that the panic in other projects never approached the scope of that engendered by World War II. An economic crisis of similar proportions might well lead to a technology-for-economic-development Manhattan Project. It is difficult to imagine any other path from the market myths that hold sway in the U.S. economic culture to the myth-exploding behavior that would necessarily accompany a dominant cooperative technology paradigm.

Scenario: Institutional Design Paradigm Dominance

Even as an analytical game it is difficult to envision the dominance of the institutional design paradigm because it is a complement rather than a stand-alone approach. So the question becomes, "What would happen were the institutional design paradigm integrated effectively with the existing three science and technology policy paradigms?" Surely the result would be less dramatic than in the previous three more extreme and exclusive scenarios. One result, interestingly, might be a greater stability as policies would perhaps be less vulnerable to capricious changes following shifts in political climate. The pragmatism, empiricism, and systemic focus of an institutional design policy framework would militate against any sweeping change rooted solely in counts of congressional Republicans and Democrats. This is not to say, of course, that federal laboratory policy would be apolitical. That is not even a desirable prospect. But it might be something other than exclusively political. If there is greater systemwide evidence for science and technology policy decision making, the evidence will, as in other policy realms, be put to diverse political uses. This is not such a bad outcome. Indeed, it is a fair working definition of rationality in a democratic, pluralistic policy-making system.

In our view as optimistic dataphiles, serious attention to an institutional design paradigm yields a U.S. R&D laboratory system more effective, better rationalized, and better able to contribute to U.S. competitiveness. The institutional design paradigm seems particularly well suited to an environment of limited resources and difficult choices. By demystifying the mysterious 16,000, policy choices, even if they remain difficult, occur in the glaring light of empirical reality.

Finally, as a summing up of summaries, we present table 8.1. In this work we posed few limits on ourselves. Hoping for absolution for the plethora of data, data summaries, typologies, and factoids we foisted upon the reader, we present a succinct summary of the data-based (and personal bias-based) recommendations it has taken fourteen years to develop. But we hope you did not begin with the last page.

Table 8.1	**Realigning Federal Labs: Policy and Institutional Recommendations**

Improved Information for Strategic Policy-Making

- Use of system-based rather than sector-based data
- Collection of panel data on all major contributors in the national innovation system

Revised Measures of Success

- Consideration of both the commercial value *and* scientific/technological activities of labs

Critical Institutional Redesign Factors

Greater Autonomy	■ Increasing lab control in technical agenda-setting
	■ Continue to give lab directors discretion over decisions about collaboration with university and industry
	■ Decrease the influence of topical program monitors
	■ Increase accountability through biannual external evaluation
Mission Rationalization	■ Zero-based assessment of mission needs, separate from considerations of capabilities and deficits
	■ Problem-directed missions
	■ Assignment of "independent agency"-like status to the best large federal labs, so these labs would have their own lines in the federal budget
	■ Increasing the number of labs designated as technology development laboratories, whose chief role is to promote industrial competitiveness and which will be evaluated by industry-based panels

Federal Lab Realignment Commission	▪ Create the Federal Lab Realignment Commission to: oversee a zero-base assessment of federal labs with a view to closing some labs; make recommendations about organizational change, redeployment of labs, and creation of new labs; and investigate new institutional designs for achieving national science and technology objectives
Coordinating Mechanisms	▪ Consideration of organizational affiliation of a lab based on the degree to which the lab's technical activity serves an agency's mission, clientele, and on the degree to which the agency "adds value" to the lab ▪ Creation of mechanisms to enhance lab-to-lab linkages
Opportunity Cost Approach	▪ Focus the assessments of multipurpose labs on the entire arrays of their technical activities and missions ▪ Commission analysis to determine the effects of commercialization activities on basic and applied research
Job Retraining and Relocation Programs	▪ Earmark funds saved from closing or cutting back federal labs for job retraining and relocation programs for displaced employees ▪ After retraining and relocation programs are fully implemented, use savings from downsizing and closures for creation of new federal or hybrid labs focusing on missions that the Federal Lab Realignment Commission determines to be "high priority"

Technology Transfer and Commercialization

Increasing Efficacy of Lab-Industry Interaction	▪ Increasing Efficacy of Lab-Industry Interaction ▪ Focus of "recruiting" efforts on companies that have a clear disposition to work with a particular lab ▪ Increased attention to determining the extent to which the federal labs' activities complement, substitute, enhance, or detract from other sources of technical knowledge
Improving Measurement of Technology Transfer and Commercialization	▪ Explicitly state models of and criterion for "effectiveness" of labs ▪ Use multiple methodologies in evaluating lab performance ▪ Commission routine low-cost, "low-confidence" evaluations of federal labs, focusing on individual projects and programs at individual labs, and periodic "high-confidence" studies of sets of labs and multiple projects and programs
Improve Commercial Effectiveness	▪ Identify which labs have the most potential for contributing to technology commercialization goals ▪ Focus on economic measures other than job creation ▪ In developing technology transfer policies, policy makers should remember that firms, like labs, are often very different from one another; some firms are more likely to be successful technology-development partners than others

Technology Transfer Strategies	▪ Federal labs should begin to "market" their direct basic and applied research ▪ Consider designating those federal labs that score highest on industrial partners' value ratings as industrial competitiveness labs, substantially increasing their commercial activities ▪ Establish a national fund to be distributed to labs that have highly rated technology transfer and commercial effectiveness
Federal Labs and Competitiveness	▪ Rationalize labs so that they have clearly defined roles ▪ Develop means to clearly communicate missions to individual labs ▪ Transform the structure of labs designated as commercial, including industry representatives on the governing boards ▪ Create legislation similar to the Intergovernmental Personnel Act to facilitate personnel exchange between commercially oriented federal labs and industry ▪ Implement a system of sector-blind grants to facilitate the commercialization of federal lab technology ▪ Promote competition among pairs of industry development labs to stimulate commercial activity ▪ Develop policies that attract the best technical and entrepreneurial talent to the federal lab system

The National Comparative Research and Development Project

Phases, Objectives, and Methods

As a result of limited empirical knowledge about the system, R&D policy making is poorly rationalized. It is difficult to even begin thinking strategically about how best to deploy national R&D resources. The institutional design perspective advocated in part 1 requires considerable knowledge of the institutions being designed or redesigned, certainly a great deal more knowledge than is typically brought to R&D policy making.

The chief resource for this monograph is the set of findings developed under the National Comparative Research and Development Project (NCRDP). Despite considerable diversity in approach and findings, the more than thirty studies (see appendix 1) performed under the NCRDP have had the same overall objectives: to provide a baseline of empirical knowledge about the R&D laboratory systems of the U.S. and other nations; to provide new conceptual tools for thinking about R&D systems and institutions. Since 1984, NCRDP researchers have obtained data from literally thousands of scientists, science administrators, and science policy makers. Further, NCRDP researchers have personally visited and studied more than 150 R&D laboratories in seven nations. As mentioned in part 1, this monograph is a data-informed policy analysis that draws from various NCRDP studies as well as from the personal experience of the researchers. Its purpose is to address some of the major strategic questions confronting policy makers concerned with the U.S. R&D laboratory system in general and federal laboratories in particular.

Part 2 presents an overview of the NCRDP. The purpose of this overview is to provide a background of the NCRDP that will offer the reader a better understanding of the nature and magnitude of the data from the NCRDP and the study procedures employed in its various phases.

To this point, there have been five distinct phases of the NCRDP:

Phase I focused on the population of 825 energy R&D laboratories, developing 32 intensive case studies (Crow and Bozeman, 1987a,b; Crow and Emmert, 1987; Crow, 1985; Bozeman and Loveless, 1987) and also involved a series of intensive project-based case studies at one DOE multiprogram laboratory (Bozeman and Fellows, 1988). In addition this phase also implemented a detailed effectiveness evaluation of the 32 case study labs.

Phase II expanded the focus to include a population of about 16,000 U.S. R&D laboratories in all fields of science and engineering, surveying data to examine a sample of 935 laboratories. (Bozeman and Crow, 1990, 1991; Crow and Bozeman, 1991; Rahm, Bozeman, and Crow, 1988)

Phase III focused on the dynamics of change, resurveying Phase II respondents, with a focus on government laboratories (Bozeman and Coker, 1992; Bozeman, in press; Coursey and Bozeman, 1992). Several technology transfer studies were produced from this phase (based on a questionnaire and lab site visits that focussed chiefly on technology transfer and cooperative R&D).

Phase IV (the only major component of the NCRDP not used for this document) focused on government laboratories in Japan and South Korea (Bozeman and Pandey, 1994; Papadakis et al., 1995; Crow and Nath, 1990, 1992).

Phase V for which a major report was written (Bozeman, Papadakis, and Coker, 1995), gathered data from more than 200 companies that interact with federal laboratories in order to determine the nature and value of those interactions.

A fundamental assumption of the NCRDP is that effective public policy for R&D requires systemic thinking about R&D laboratories. Typically, R&D laboratories are examined either individually, by sector, by industry, or by product attributes. Rarely is there sufficient consideration of R&D performers as a system, as a set of interacting components encompassed by boundaries and constrained by resource needs and other identifiable interdependencies. Indeed, several studies from both the first and second phases of the NCRDP have been concerned with conceptualizing the U.S. R&D system, including system profiles, developing taxonomies for classification of laboratories, and both generating and testing propositions about the relationship of laboratory types to environments.

The Goal of the NCRDP: Knowledge of the U.S. Laboratories as a "Knowledge Production System"

The current era in U.S. science and technology policy is one of rethinking and redirection. Some of the more prominent changes include:

- an emphasis on centralization of R&D and on the design of science and technology centers,
- a focus on interdisciplinary research,
- linkage of industrial, university, and government R&D through new institutions and cooperative R&D missions,
- a refocusing and possibly complete redesigning of government laboratories, especially the so-called national laboratories,
- an effort to more quickly and thoroughly appropriate to commercial purpose the output of government laboratories and government-sponsored R&D

Each of these developments has the potential to profoundly change the structure and activities of actors within the U.S. R&D laboratory system. But without greater knowledge of that system, its elements, and its dynamics, it is difficult to evaluate or even document change. Although policy makers bring a storehouse of hunches, well-worn stereotypes, and some valid historical knowledge to bear on policies directed at institutional change in R&D, there is little evidence of system effects due to the lack of knowledge about the system.

With better system-level knowledge many policy relevant issues might be elucidated. Some of these include:

1. What are the major categories of R&D laboratories operating in the 1990s?
2. What are the various categories of information and technology laboratories produce and what is the value of these products?
3. What are the structural and environmental characteristics of each laboratory category?
4. How do the various laboratory types respond to public policy initiatives?
5. Given certain policy or system goals, what laboratory types are best suited for production?
6. Are there weaknesses in the system?

Phases of the NCRDP

From relatively small-scale beginnings in 1984—a study based predominantly on 250 surveyed labs and on intensive case studies of 32 laboratories devoted to energy research—the latter stages of the NCRDP have examined more than 1,000 laboratories using multiple data sources and encompassing the full spectrum of institutions contributing knowledge, technology products, and technical assistance to the technology development and innovation process. Each of the phases is described below.

The NCRDP was not planned as a multistage research project. Thus, incremental decisions about continuing gaps in our knowledge rather than in one systematic mapping out of a research agenda determined its evolution. However, each of the phases addressed a distinct problem.

NCRDP Phase I: Case Studies of Energy R&D Laboratories

The first phase of the NCRDP was primarily designed for the purpose of developing and testing empirically a taxonomy of R&D laboratories based not on traditional sector-based (i.e., industry, government, university) distinctions but on the impact of two fundamental features of laboratories' environments: the influence of the government and/or private resources base and the market orientation of the laboratories' R&D products as either public domain or proprietary. (The study procedures and results of Phase I are reported in Crow and Bozeman, 1987a; Crow and Bozeman, 1987b; Crow, 1985, 1987; Crow and Emmert, 1987; Emmert and Crow, 1988)

Phase I was based on an interest in examining the effects of hybrid organizations (partly government, partly private) and "publicness" (Bozeman, 1987). The chief idea behind the dimensional model of publicness is that "publicness is not a discrete quality but a multidimensional property. An organization is public to the extent it exerts or is constrained by public authority" (Bozeman, 1987). It was assumed that R&D laboratories were an excellent choice for the analysis of government authority and market interactions. A number of studies were undertaken that, among other objectives, attempted to determine the impact of government involvement (via resources, agenda-setting, procedural requirements) on R&D laboratories, including private sector labs. At the same time, a parallel effort (Bozeman and Loveless, 1987; Loveless, 1985) provided a straightforward sector-based comparison of the performance of both government and industry laboratories. The two sets of studies, when taken together, indicated that both ownership (government, business, university) and "publicness" (i.e., government constraint) had important and independent effects on the laboratories' structure and performance.

Phase I: Methods and Procedures
The data base for Phase I was derived from the population of all U.S. and Canadian R&D laboratories engaged in energy-related research and development.

Using laboratory directories (Cattell Press, 1970, 1983) and personal telephone calls, some 829 energy-related R&D laboratories were identified. While the chief focus of NCRDP Phase I was case study analysis, a survey was conducted for the purpose of developing and testing a classification taxonomy. During March and April 1984, a questionnaire was mailed to the directors of each of the 829 laboratories. After sixty days had elapsed and both follow-up letters and phone calls had been implemented when necessary, a respondent pool of 250 usable surveys had been compiled.

The case studies involved site visit to 32 R&D laboratories in the U.S. and Canada. In each case, a common interview format was followed in a meeting with the laboratory directorate (the director and members of the senior management staff). The objective of the fieldwork included analysis of laboratory history, organizational structure, resource flows, types of research conducted, planning procedures and formats, and scope of research. Laboratory tours and extensive site visits were conducted in each case.

Overview of Findings from Phase I

The findings from Phase I (as well as other Phases of the NCRDP) are provided throughout this monograph. Nonetheless, it is useful to provide here a list of general findings.

As a result of the 32 case studies and the attendant questionnaire data ($n = 250$), a taxonomy was generated that examined laboratories in terms of their mix of market and political authority (as opposed to sector or ownership) (Crow and Bozeman, 1987a; Crow, 1985). The "publicness" of R&D laboratories was defined in terms of the level of government funding ("High publicness" labs had 76–100 % of their funding from government; "Moderate publicness" had 26–75%; "Low publicness" had 0–25%). The "market influence" of labs was defined in terms of the economic nature of their primary R&D products, including generic products (e.g., knowledge products in the public domain), balanced products (knowledge products and market-directed products), and proprietary products (e.g., production goods and technology). Out of these two dimensions—market influence and public influence—a typology was created (see table I.1). This typology served as the rough draft for what would become the "Environmental Context Taxonomy," a conceptual tool used in a number of subsequent NCRDP studies (Bozeman and Crow, 1991a, 1990, 1988).

The question, however, was whether the laboratories in the study actually conformed to the typology and whether the typology predicted lab behavior (in the sense that laboratories of a similar category in the typology tended to behave similarly). As is indicated in tables II.2, II.3, and II.4 similar tendencies were found within each of the laboratory types. Some specific findings included:

Table I.1	Environmental Context Taxonomy		
Level of Market Influence	Level of Government Influence		
	Low	Moderate	High
Low	Private Niche Science	Hybrid Science	Public Science
Moderate	Private Science and Technology	Hybrid Science and Technology	Public Science and Technology
High	Private Technology	Hybrid Technology	Public Technology

- production of generic (non-market-directed) research products requires a stable environment and either high levels of publicness (i.e., higher percentage of government funding) or a balance of government and market influences;
- higher levels of publicness are associated with longer planning horizons and thus a much longer term agenda and decision making environment than lower levels of publicness;
- proprietary R&D environments have short-range planning horizons due to the need to adjust implementation to market changes;
- goal conflict (between market and government influences) results in highly unstable R&D environments;
- organizational structure is not related to publicness but rather to the nature of the R&D product: the more proprietary the product, the more centralized the structure.

The most important development of Phase I analysis was simply demonstrating that something is to be gained by developing new typologies and categories for thinking about R&D laboratories. The implications of these approaches were more fully realized in Phase II.

The Phase I analysis touched on the following theme: the very different requirements of "technology transfer" (i.e., proprietary products) versus "knowledge transfer" (i.e., generic products), which has subsequently influenced much of the NCRDP work. Another major product of Phase I (Bozeman and Fellows, 1988) was based on a set of extensive case studies performed at Brookhaven National Laboratory. While the chief purpose of these case studies was to develop models for evaluating the effectiveness of technology transfer, the cases made clear that knowledge transfer projects require quite different assessment approaches than technology transfer. The case included assessment approaches to the transfer of a physical technology (superconducting magnet), a technological process (polymer insulation), "know-how" (remotely operated robotic arm), resource sharing and brokering (studies from the National Syncrotron Light Source), and technical assistance (Three Mile Island corrosion research). The cases resulted in a framework (Bozeman and Fellows, 1988) for assessing technology and knowledge transfer activities that was based on four different concepts of effectiveness: "out-the-door" (i.e., something was transferred to another organization), "market" (the transfer object was brought to market), "opportunity costs" (considers possible alternative uses of resources employed in technology transfer), and "political" (aims at developing increased lab resources through political support). This framework and the associated effectiveness models were employed in a variety of subsequent NCRDP studies (e.g., Bozeman and Coker; 1992, Bozeman and Pandey, 1994)

Table I.2   General Laboratory Characteristics

Level of Market Influence	Level of Government Influence		
	Low	**Moderate**	**High**
Low	**Private Niche Science** N = 6 A = 1953 F = 0.753 GI = 3.53 OBJ = 4.4	**Hybrid Science** N = 40 A = 1957 F = 1.2 GI = 2.52 OBJ = 5.3	**Public Science** N = 121 A = 1958 F = 2.95 GI = 2.13 OBJ = 5.68
Moderate	**Private Science and Technology** N = 10 A = 1940 F = 8.4 GI = 4.8 OBJ = 2.1	**Hybrid Science and Technology** N = 11 A = 1958 F = 2.0 GI = 3.27 OBJ = 3.6	**Public Science and Technology** N = 4 A = 1961 F = .692 GI = 2.29 OBJ = 5.75
High	**Private Technology** N = 30 A = 1953 F = 6.795 GI = 4.75 OBJ = 1.63	**Hybrid Technology** N = 11 A = 1957 F = 1.5 GI = 3.18 OBJ = 2.18	**Public Technology** N = 10 A = 1959 F = 5.65 GI = 2.5 OBJ = 4.3

Key:
N = Number of Labs
A = Date of startup (Average)
F = Funding level in millions of dollars (1985) (Mean)
GI = Level of Government Influence (5 = low & 1 = high) (Mean)
OBJ. = Objective of research (1 = Commercial & 6 = Public) (Mean)

Table I.3   **Organizational Variation of R&D Organizations**

	Public Science	Hybrid Science	Private Science	Public S&T	Hybrid S&T	Private S&T	Public Technology	Hybrid Technology	Private Technology
Research scope	National interfirm	Regional intrafirm	Regional intrafirm	Regional intrafirm	National interfirm	Regional intrafirm	Regional intrafirm	Regional intrafirm	Regional intrafirm
Type of technical change pursued	Revolutionary	Evolutionary	Evolutionary	Evolutionary	Revolutionary	Evolutionary	Evolutionary	Evolutionary	Evolutionary and sometimes revolutionary
Research frontier	Scientific and technological	Achievement and technological	Scientific and technological	Technological	Achievement and technological	Achievement	Achievement	Achievement	Achievement and technological
Technological research planning	Multiyear	Annual to Biannual	Annual to Biannual	Biannual	Annual to Biannual	Annual	Annual to Biannual	Annual	Annual
Organizational stability	Stable	Unstable	Unstable	Unstable	Stable	Unstable	Unstable	Unstable	Stable
Organizational structure	PI Mode	PI Mode	PI Mode	Departmental	Departmental and PI Mode	Departmental	Departmental	Departmental	Departmental

| Table I.4 | | **Laboratories' Scientific and Technical Output and Means, by Sector and Taxonomic Category** | | | | | | |

Ownership	Scientific articles	Proto-types	Demonst.	Patent	Reports	Paper	Algo-rithms
Government	36.4	10.3	8.5	2.1	16.7	12.9	8.2
Industry	4.0	32.6	9.6	5.3	20.0	4.4	8.2
University	44.6	6.3	4.0	2.1	10.5	18.8	7.0
Other	55.8	5.8	3.0	2.6	18.5	13.5	5.0
Taxonomic type							
Hybrid science	42.9	4.6	1.2	1.7	12.1	21.0	5.0
Public science	46.3	5.9	4.7	1.4	13.1	16.4	7.3
Private science and technology	5.4	25.0	7.6	5.9	17.1	5.2	7.7
Hybrid science and technology	16.3	22.3	9.6	4.9	19.1	7.6	10.7
Public science and technology	27.1	16.5	11.4	3.2	12.3	12.4	10.5
Private technology	2.4	36.0	8.8	6.2	23.6	3.5	4.7
Hybrid technology	3.2	37.1	9.6	4.5	19.6	3.7	9.6
Public technology	4.5	26.7	13.3	1.6	16.0	5.1	16.4

NCRDP Phase II: Profiling the U.S. R&D Laboratory System

While Phase I of the NCRDP seemed to demonstrate the need for new ways of thinking about R&D laboratory environments, there were important limits to this early work. First, and most obviously, to what extent could energy R&D laboratories be viewed as representative of all R&D laboratories? This was particularly troubling because energy laboratories typically have greater entanglement with government (especially during the 1970s and early 1980s) and among the population of energy laboratories there are more "hybrids" not easily classified by the usual sector categories. Another important limitation of the Phase I research was that it was more interested in establishing the taxonomy rather than in using it to predict laboratory behaviors. Thus, the survey data used in building the taxonomy had only limited utility for determining the predictive value of the taxonomy.

Phase II: Methods and Procedures

The second phase of the NCRDP aimed at nothing less than developing an understanding of the entire U.S. R&D laboratory "system." Thus, there was a concern about developing a more representative sample of U.S. R&D

laboratories and going beyond the few attributes examined in Phase I. Thus, Phase II gathered not only sufficient information to refine the Environmental Context Taxonomy but also information on a wide variety of laboratory attributes including:

- Laboratory missions
- Budgets and sources of funds
- Organizational structures
- Approaches to laboratory evaluation
- Composition of output
- Personnel characteristics
- Responses to public policy initiatives
- Interaction with government agencies

The more intensive data collection involved in Phase II also permitted further development and refinement of the Environmental Context Taxonomy and the use of the taxonomy to predict variance in each of the laboratory attributes and behaviors listed above.

The data reported in Phase II were derived from questionnaires, both mailed and administered by phone.

In Phase II, four major research center directories were used to establish a population of U.S. R&D laboratories. Laboratories with less than 25 reported employees were excluded from the study population as were those chiefly conducting research in the social sciences. This yielded a study population of 16,597 R&D laboratories.

In drawing the sample for this study, both random probability and stratified sampling were used. A random probability sample of 1,300 labs was developed using a computer-generated random number list. In addition to random probability sampling, to assure representativeness, it was deemed useful to gather information about the largest R&D laboratories in the U.S. Since the researchers were interested in ensuring statistical significance at the <.01 level for a two-tailed test, a list of 1,300 was drawn for the sample. The largest 200 laboratories (as determined from analysis of total laboratory personnel figures) were added to this list. It was anticipated that a response rate of about 40% would be both feasible and suitable for the purposes of the study.

The researchers recognized that the data provided in the most recent standard research directories would necessarily be somewhat out of date and would entail at least a few coding and other errors. To compensate for these problems, each of the 1,500 laboratories were telephoned by the researchers in order to confirm the continued existence of the laboratory, correct addresses, develop data about areas of research focus and total personnel, and to confirm the name of the current laboratory director. As a result of this process, the study sample was reduced from 1,500 to 1,341.

The design of the questionnaires was undertaken jointly by the researchers. At the outset, it was decided that most of the questionnaire items should be a discrete item in nature, that the length of the mailed questionnaire should be less than twelve pages, and that a mix of objective and opinion data would be elicited. Beginning with previous theoretical frameworks, related previous studies, and explicit hypotheses, a master list of questionnaire items was developed. The researchers agreed on priorities among the questionnaire items and developed an instrument for pretest.

A separate random probability sample of 60 was drawn from the population by identical computer-generated random number techniques. In addition, to indicate the response patterns for the 200 largest R&D laboratories (the "superlaboratories"), a group of the next 20 largest labs (201–221) was included in the pretest. As with the more general sample, research assistants telephoned each of the firms to ensure correct addresses and to double-check the name and continued tenure of the laboratory directors who comprised the intended respondent pool. The approach of the more general study was used to the extent possible. From the 80 questionnaires mailed, 31 usable questionnaires were returned. The researchers analyzed the responses in order to determine possible ambiguities, the degree of response variation, and to compare known characteristics of the respondents to known characteristics of the population, checking for degrees of nonresponse bias. From this information, the questionnaire was revised again.

After considering the results from the pretest, it was clear that not all of the desired information could by obtained from the mailed questionnaire. The necessary length for a questionnaire including all the desired items would have been prohibitive. Because of the desire for additional information and the concerns about the likely difficulty of obtaining a response rate of the desired 45–50% from the mailed questionnaire, a telephone questionnaire was developed. The telephone questionnaire included questions from early drafts of the questionnaire, but often revised in scale for convenience of administration. Telephone calls were completed to 1,012 laboratory directors. Among these directors, 88 were deemed inappropriate as respondents for the study (not meeting one or more of the criteria pertaining to size and focus of the laboratory) and 665 participated for a response rate of 71%. Of the 1,341 eligible laboratories contacted by phone and questionnaire, data were received (phone and/or questionnaire) from 966 for an overall response rate of 72%. Considering just the mailed questionnaire, 711 usable responses were received for a response rate of 53%.

OVERVIEW OF FINDINGS FROM PHASE II

Phase two research papers dealt with diverse topics, including levels of bureaucratization and red tape (Bozeman and Crow, 1991b; Bozeman, Reed, and Scott, 1992), technology transfer (Rahm, Bozeman, and Crow, 1988), and the relation of firm size to innovation (Link and Bozeman, 1991), among others. These were particularly important for fleshing out the Environmental Context Taxonomy. The implications of the taxonomy are not conveniently summarized (see chapter 4 for detailed discussion of taxonomic work). But table I.5 provides an indication of the distribution and incidence of each laboratory according to sector (government, industry, university, other). Without dwelling on the full implications of each laboratory type, at this point, that table (from Bozeman and Crow, 1990) makes clear that laboratories of different sectors are widely distributed by type. Thus, for example, market forces clearly influence government labs and political forces clearly influence industry labs.

The taxonomic work also produced less complex findings related to the effects of government influence versus market influence (apart from the

Table I.5   Level of Market Influence

Level of Market Influence	Level of Government Influence		
	Low	**Moderate**	**High**
Low	**Private Niche Science**	**Hybrid Science**	**Public Science**
	Not available	Total [54]	Total [133]
		Government 2	Government 52
		Industry 6	Industry 10
		University 33	University 63
		Other 13	Other 8
Moderate	**Private Science and Technology**	**Hybrid Science and Technology**	**Public Science and Technology**
	Total [59]	Total [73]	Total [77]
	Government 0	Government 0	Government 37
	Industry 53	Industry 53	Industry 22
	University 3	University 13	University 14
	Other 3	Other 7	Other 4
High	**Private Technology**	**Hybrid Technology**	**Public Technology**
	Total [175]	Total [82]	Total [25]
	Government 0	Government 1	Government 5
	Industry 170	Industry 76	Industry 17
	University 4	University 4	University 3
	Other 1	Other 1	Other 0

sector of the labs). Phase II research (Crow and Bozeman, 1991) indicated that the following attributes are associated with "publicness":

- More basic research
- More cooperative research
- Greater red tape and bureaucratization
- Stronger emphasis on technology transfer
- Greater focus on scientific effectiveness
- Policy-driven instead of market-driven
- Larger size
- Shortage of scientific personnel
- Higher levels of interorganizational complexity
- Great mix of knowledge and technology products

By contrast, increased market influence (regardless of sector) on laboratories tends to have the following impacts:

- Focus on applied research, with little basic research
- Slower time for release of new knowledge
- Concentration on engineering
- Less interdisciplinary research
- Smaller R&D environment with less interorganizational interaction

In addition to the findings related to the taxonomy, Phase II studies provided some useful information about bureaucratization, red tape, and the structure of R&D labs (Crow and Bozeman, 1989a; Bozeman and Crow, 1991a). One innovation was developing a behaviorally based measure of "red tape" as the amount of time taken to provide core functions of the organization, such as hiring full-time personnel, buying equipment, circulating research results, and obtaining approval for research projects. As indicated by the NCRDP (Crow and Bozeman, 1989a), government labs tended to have higher levels of red tape (defined by weeks to perform tasks) than labs of other sector types. Subsequent studies (Bozeman, Reed, and Scott, 1992) indicated that, unlike many attributes of laboratories, the sector status (ownership) was more important in determining levels of red tape than was publicness (i.e., government funding).

NCRDP Phase III: Government Labs and Technology Transfer

The third phase of the NCRDP was completed in 1992. Despite the amount of information generated from earlier studies there was a clear limitation—the picture was a static one. The third phase of the NCRDP was designed to permit some analysis of change. A subset of laboratories examined in 1984–1988 was surveyed again in 1990–1991 for the purpose of understanding some of the dynamics of change. Phase III differs from the earlier phases in two other respects. Because of our interest in learning more about government laboratories, survey questionnaires were sent to *every* government laboratory that met our criteria for analysis. The last several years have brought great change in the entire R&D system, but the government laboratory component has been especially affected by the policy changes of the 1980s. Another major theme of the past decade has been an increased emphasis on technology transfer and cooperative R&D. Thus, Phase III gave particular attention to those rapidly evolving issues and policies.

PHASE III: *Methods and Procedures*

Phase III analysis was based on responses to questionnaires mailed to laboratory directors. Between June 1990 and August 1990, questionnaires were mailed to each of the laboratory directors who had participated in the Phase II study as well as to directors of all government laboratories meeting the following criteria: (1) focus on science and engineering rather than social science; (2) more than 30 total personnel. Designed as a panel study, Phase III sought data from all government labs, all respondents from Phase II, and focused intensively on technology transfer and cooperative R&D.

The Phase III sample consisted of 1,137 laboratories. A total of 533 questionnaires were returned for an overall response rate of 47%. By sector, questionnaires were sent to 594 industry labs (260 received, 44% response rate); 164 university laboratories (71 received, 43% response rate); 23 nonprofit or hybrid laboratories (12 received, 61% response rate); and 356 government laboratories (189 received, 53% response rate). Given a concern to measure change, most of the sample (939 of the 1,137) and most of the respondents (420 of the 533) were drawn from the pool of respondents to a 1988 Phase II questionnaire.

OVERVIEW OF PHASE III FINDINGS

Most of the research published from Phase III focused particularly on government laboratories and technology transfer. One study linked a Phase II theme, bureaucratization and red tape, to the two major themes of the more recent phase, namely red tape in the technology transfer of government labs. This study tested a series of regression models, regressing measures of technology transfer success derived from earlier conceptual work (Bozeman and Fellows, 1988) on a variety of red tape measures, including not only the delays described above but also perceptual measures and the number of administrative levels. The most important finding was that technology transfer activity was *not* associated with increased levels of red tape (by any measure). Other findings included:

- Government laboratories with less red tape have higher effectiveness levels in transferring technology.
- Government laboratories have significantly more red tape, both behavioral and perceptual, than industry laboratories.
- Government laboratories and industry labs differ little in their administrative levels and spans of control.
- "Out-the-door" technology transfer success is most closely related to low levels of perceived red tape and less time required for approval of small projects.
- "Market impact" technology transfer success is most closely associated with less time required for the approval of large projects and buying of equipment.

Much of the technology transfer research from the NCRDP uses the Phase III data. The number of technology transfer and cooperative R&D-related questions was greatly increased in the Phase III questionnaire. The technology transfer findings are reviewed in detail in part 5 of this report, but it is worth summarizing at this point some of the chief findings. Two of the major technology transfer studies from Phase III (Bozeman and Coker, 1992; Bozeman, in press) shared measures of technology transfer success, including the numbers of licenses and patents, the responses to questionnaire items providing lab directors' self-assessments of "out-the-door" success and market impact success of laboratory technology transfer.

Some of the more interesting findings were purely descriptive. For example, the federal laboratories had produced relatively few technology licenses and just a handful of labs accounted for most of the licensing activities. Other findings included:

- Multifaceted, multimission laboratories are likely to achieve greater technology transfer success
- Technology transfer effectiveness is related to low levels of bureaucratization.
- "Market success" and "out-the-door" success are not strong correlates.

In a more comprehensive Phase III study (Bozeman, in press) of technology transfer activities, looking not only at effectiveness correlates but also at motives, strategies, and organization for technology transfer, the following findings emerged:

- Among several technology transfer strategies, the most effective seems to be participation in a research center. Close behind is use of cooperative R&D for technology transfer.
- The strategy of using a central office for technology transfer doesn't have much effect *except* that it enhances the likelihood of pecuniary benefit to the lab's scientists and engineers.
- Regarding the motivation of labs for technology transfer, an emphasis on economic development is clearly the most closely tied to success.
- Some laboratory missions complement technology transfer success, others don't. Specifically, a focus on applied research—either commercial or precommercial—is salutary. Involvement in technical assistance is related to technology transfer effectiveness, but negatively if the target of assistance is the parent agency, positively if the target is industry.
- There is only a modest relation between the organization of work (department, principal investigator, and so forth) and technology transfer success, but labs with a variety of organizational schemes appear to be more successful.
- One of the strongest correlates to success in licensing is simply the size of the laboratory, whether measured in total personnel or size of budget.

These fairly detailed findings from Phase III are derived from data collected from government laboratories. In part 6 of this report, these findings are integrated with, and sometimes contrasted with, Phase V data collected from federal labs' industrial partners.

Phase IV: Internationalization of the NCRDP

The purpose of Phase IV of the NCRDP, undertaken between 1990 and 1994, was to develop data from other nations similar to the data developed for U.S. R&D laboratories. The chief focus of this work has been on Japan's government laboratories (Bozeman and Pandey, 1994; Papadakis et al., 1995). With the cooperation of the National Institute for Science and Technology Policy (a division of Japan's Science and Technology Agency), a Japanese language questionnaire virtually identical to the Phase III NCRDP questionnaire was developed. This questionnaire was provided to directors of all of the Japanese government R&D laboratories and, with NISTEP assistance, all but two responded with usable questionnaires. In addition, more than 30 of these labs were visited by the principal investigators of the project, and laboratory directors were asked questions about approaches to setting their research agendas, basic research initiatives, research evaluation techniques, and cooperative R&D. Questionnaires similar to the base Phase III questionnaire have also been developed and deployed in research on Korean R&D laboratories (Crow and Nath, 1992) and, most recently, in Canada (Niosi, 1994). In addition, special studies of Japanese industrial research (Crow and Nath, 1990) and comparative assessments of U.S. and Japanese system response to basic science breakthroughs (Crow, 1989) were carried out.

Phase IV studies are not examined as part of this monograph although they obviously influence our thinking.

Phase V: Industry Perspectives on Federal Laboratory Interactions

Phase V of the NCRDP is concerned exclusively with federal laboratories, specifically the benefits of industry's interaction with federal laboratories. The source of Phase V data is a questionnaire mailed to companies interacting with federal laboratories in such modalities as cooperative R&D (including CRADAs), technical assistance, technology transfer, licensing, and R&D consortia. Phase V is particularly relevant to the current monograph and is discussed in detail in chapter 5. The only product from Phase V (which is still underway) is a final report (Bozeman, Papadakis, and Coker, 1995b) to the sponsor, the National Science Foundation.

PHASE V: METHODS AND PROCEDURES

A distinctive feature of Phase V is the focus on interactions between federal laboratories and private (usually industrial) organizations. In the interest of casting a wide net, the study focused not on any particular type of commercial interaction, such as a CRADA or license, but on all types of interactions.

Selection of Federal Laboratories for the Study

It was decided, then, to focus on "commercially active" federal laboratories and their interactions with industry. The problem with this approach was determining ahead of time which labs should be included in the commercially active set. The problem of identifying commercially active labs was greatly alleviated as a result of the cooperation of the U.S. GAO, whose officials provided access to a study of the more than 300 federal laboratories' commercial activities. After examining the GAO data, the researchers decided to include laboratories in the commercially active set if they met any *one* of the following criteria:

12 personnel visiting from industry
6 CRADAs with U.S. business

8 total patents issued by the lab

8 patents issued

or $50,000 in royalty income

Using these not particularly stringent criteria, only 54 (of the more than 300 larger (labs with 225 personnel) federal laboratories could be considered commercially active.

We did not include all the commercially active federal laboratories in this study. This was considered unnecessary since our focus was on the interaction rather than on the specific laboratory. Most of the laboratories did agree to comply with our requests for data. Industrial partners and projects do represent interactions with some (or all) labs from the following agencies: DOE, DOD, Bureau of Mines (Department of the Interior), NASA, CDC, the FDA, and NIH (Department of Health and Human Services). Although the technology transfer offices of Agriculture and NIST willingly provided data, their information that could be released to the public was not sufficient to be used in this research.

Table I.6 presents several basic elements of the preliminary sample frame. The labs that are included in this study are identified, as well as the scope of technical interactions that the projects/partners represent, the initial number of projects/partners provided, and the number of those actually surveyed.

The initial research design anticipated that the commercially active labs would be able to provide detailed information on their industrial partners, including project names, company contacts, and company addresses. Discussions with the labs through letters and phone calls revealed that this was not the case for three particular reasons. First, many labs do not keep any centralized records of this sort, and the appropriate data cannot be readily compiled. Second, because of confidentiality policies, several labs (or their parent agency) would not provide the needed information (information that was also immune to Freedom of Information Act [FOIA] requests). Third, these sorts of "technology transfer records" were often kept and provided by the parent agency without distinguishing among individual laboratories.

SAMPLE FRAME

Not all of the industrial projects and partners that the labs provided were actually surveyed. In some instances, specific projects/partners were outside the scope of the research design (e.g., university partners). In others, insufficient information was available. The following criteria were used to retain a project or industrial partner within the sample population:

- Universities and public agencies were dropped, retaining only private sector organizations.

| Table I.6 | **Preliminary Sampling Frame: Federal Labs' Interactions and Projects Examined** | | |

Lab/agency name	Type of Interactions	Total number of Projects	Number Surveyed
DOE Multiprogram Labs Oak Ridge Idaho Engineering Argonne Lawrence Livermore Lawrence Berkeley Los Alamos Brookhaven Sandia	All interactions	580	423 plus 80 in the pretest
Naval Research Lab (DOD)	CRADAs and licenses	27	13
Rome Lab (DOD/Air Force)	CRADAs and licenses	29	26
NASA: all labs	Officially documented spinoffs	94	
Pittsburgh Research Center (BLM)	All interactions	631	random sample of 125; 71 surveys
HHS Labs: All NIH Labs All CDC Labs All FDA Labs	CRADAs and licenses	531	random sample of 109; 74 surveys

- A specific project title had to be available and identified.
- A specific project manager or company contact had to be identified or located through phone calls to the company.
- Full mailing addresses had to be identified or located.
- A project manager could not receive more than one survey.

Questionnaires and other project-related correspondence were mailed to the individual identified as the project head or point of contact at the firm. The questionnaires asked for detailed knowledge regarding *specifically identified* projects.

SURVEY PROTOCOL AND RESPONSE

After a pretest and revisions, a final questionnaire was sent out in stages between February and April 1994.

In all, the researchers mailed 694 questionnaires to companies that were known to have technical interactions with the federal laboratories. Of those 694, 24 questionnaires were returned as undeliverable, 16 had disconnected phone numbers, and 53 respondents provided information indicating that the project should not be included in the study, either because it was not a commercial interaction or because it did not involve a private organization. This yielded an effective response rate of 38.3%. While this response rate is somewhat lower than that of related studies of the U.S. federal laboratory system (Bozeman and Crow, 1988; Bozeman and Crow, 1990), it is about the same as the response rate received in earlier studies from industry lab participants.

The data collection was officially closed on May 15, 1994. A draft final report was completed in August 1994, and the final report was turned in to the NSF in November 1994. In July 1994 the data were transformed into a form suitable for analysis on SPSS-For Windows, from which the statistics presented here were constructed.

OVERVIEW OF PHASE V: FINDINGS

Phase V findings (with the exception of the new analysis presented in part 5 of this report) are from the NSF final report (Bozeman, Papadakis, and Coker, 1995b) and include the following:

1. Companies were asked to indicate the type of R&D or technical activity performed by the respective partners. While many of the projects (120) involved basic research performed by a federal lab, 83 involved company-performed basic research. This is somewhat surprising since only about 11% of industrial labs are significantly involved in basic research (Bozeman and Crow, 1988).

2. The most common initiators of the projects were not federal laboratory personnel but company R&D managers (111) and company bench-level scientists and engineers. What this seems to indicate is that, despite recent exhortation, the laboratories are not as proactive as they might be.

3. When asked about the reasons important "in your company's decision to work with the federal lab on this project," the most common response (127) was that of the skills and knowledge of federal lab scientists and engineers. Almost as important (105) were the unique facilities of the lab.

4. There are three major objectives for companies' interaction with the federal laboratories: engaging in strategic precommercial research (111), interest in access to unique lab expertise (98), and desire to develop new products and services (90).

5. The firms collaborating with federal laboratories in order to enhance their precommercial R&D (but *not* for product development) are smaller but

highly R&D intensive firms. Firms motivated by product development (but *not* precommercial R&D) are larger than those that focus on precommercial R&D, and they are not particularly R&D intensive.

6. Comparing firms chiefly motivated (in their interaction with a federal laboratory) by new product development and those for whom other motives are more important, the ones motivated by new products report benefits only about one-tenth of those of all other companies.

7. Barriers to technology transfer and commercialization activity are not substantial. More respondents view labs' internal management procedures as positive than as negative. The "speed with which the lab was able to take administrative actions" receives unfavorable assessment with a slight majority giving a negative assessment.

8. At the time the industrial respondents were queried, 47 of the interactions had already led to a product being marketed (18.5%) and in 79 cases products were in development. Existing products had been improved in 52 cases. In only 18 cases had product development efforts failed. This indicates that lab-industry interactions are very much driven by commercial aspirations, and just as important, that those aspirations often are met.

9. The set of companies having marketed products as a result of their interactions with a federal laboratory is both younger and smaller—and substantially so—than the other firms in the data set.

10. Commercialized products were more likely to result from projects if the projects were initiated by either the companies' R&D managers or by top managers in the company. Projects developed by bench-level scientists, lab directors or, in most instances, federal laboratory personnel, were no more (or less) likely to lead to commercial results. Projects initiated by federal laboratories' technology transfer staff were somewhat more likely to be commercialized.

11. The nature of the technical activity is *not* a good predictor of commercial results. One might expect that basic research activity would be negatively related to commercial product results (it isn't) and that development activity would be positively related (it isn't).

12. While one might anticipate that companies that develop commercial products would reap more benefit from their interactions with federal laboratories, the case is just the opposite. The findings clearly suggest that interactions resulting in commercial products are *less* beneficial than others. The results indicate that companies that had *not* yet developed a commercial product or service reported more than four times the benefits received and only 74% of the cost. Upon reflection, the latter figure is not surprising—developing products might well be expected to lead to greater cost. But what could explain the strong tendency for those with commercialized products to report *less* benefit? The most likely explanation is that the federal labs' true value as an industrial partner may be in upstream rather than downstream

work. The benefit-cost ratios are actually slightly negative ($476K to $536K) among the product development interactions while highly positive, indeed 4 to 1, on the nonproduct interactions.

13. The results indicate that the federal lab customers are, for the most part, satisfied customers. According to a global measure of satisfaction with the interaction, 89% agree or strongly agree that they were satisfied.

14. The small minority (11%) of *dissatisfied* customers had the following characteristics:

> Interaction based on a license
> Company scientists at the lab
> Lab scientists at the company
> Cost of the interaction
> Employees hired
> Employees fired
> Failed in new product development
> Previous relationship was as a federal lab employee
> Previous relationship was as a federal lab contractor
> Company R&D managers began the project

15. On average, 1.6 new jobs were created by interactions between federal laboratories and industry (which must be weighed against the .05 jobs displaced as a result of the interactions). Fully 90% of the projects do *not* result in new hires. Of all the projects examined, only three resulted in hires of more than 10 new employees.

Perhaps the most important aspect of Phase V research has been its focus on developing numerical assessments of the value of company-federal laboratory interactions. Part 5 of this report deals extensively with this issue, and thus the findings are not detailed here. In summary, however, the average level of benefit reported was $1,604,564. The average level of cost was $417,010. This finding does not indicate that the companies had a 4:1 multiplier for their interactions with the federal laboratories because many of the companies have essentially no cost for the interaction. Looking only at those companies that sustained costs for the interaction, the mean rate of return was 2.67; that is, they valued the benefits at a rate of 2.67 greater than the amount expended as company cost.

Synthesizing the NCRDP

The purpose of part 2 was to provide a broad overview of the NCRDP, placing the literature used subsequently in this monograph in a historical and methodological context. While each section presented an overview of the findings from the NCRDP, those findings only provide a snapshot of the more than 30 research papers, books, and dissertations produced during the more than ten years of the NCRDP.

What remains unclear from the foregoing is the extent to which the quantitative data of the NCRDP has been buttressed by site visits, case analyses, and interviews. Each of the questionnaires utilized in the NCRDP was preceded by considerable face-to-face interaction with laboratory personnel, government officials, industry technical and managerial personnel, and users of R&D products. The authors of this monograph have personally visited more than 200 laboratories in eight nations. Many other such site visits have been undertaken by other NCRDP participants. While some of this contextual knowledge has been integrated into NCRDP publications (e.g., Crow and Bozeman, 1987a; Bozeman and Fellows, 1988), much of it has not. It is nevertheless important to note that many of the recommendations presented in this monograph are drawn not only from the direct empirical findings from the NCRDP quantitative studies but also from the many supportive activities including site visits, case studies, and conversations with persons working in and knowledgeable about the U.S. R&D laboratory system. Thus, this monograph is, indeed, a synthesis of the NCRDP, not only its published studies but, more broadly, its accumulated knowledge.

Blue-Ribbon Panel Reports on the National Labs

1975 Report of the Field and Laboratory Utilization Study Group   DOE

1978 Report to Congress: The Multiprogram Laboratories: A National Resource for Nonnuclear Energy Research, Development, and Demonstration   DOE

1981 Federal Energy R&D Priorities   DOE

1982 The Department of Energy Multiprogram Laboratories: The Final Report of the Multiprogram Laboratory Panel, Energy Research Advisory Board   DOE

1983 The Federal Role in Energy Research and Development: A Report of the Energy Research Advisory Board to the United States Department of Energy   DOE

1983 Report to the White House Science Council   White House   OSTP

1985 Guidelines for DOE Long-Term Civilian Research and Development   DOE

1986 The Coordination of Long-Term Energy Research and Development Planning   DOE

1990 The Future Strategic Role of the Department of Energy's Multiprogram Research Laboratories   DOE

1991 Nuclear Weapons Complex Reconfiguration Study   DOE

1992 Report to the Secretary on the DOE National Laboratories   DOE

1993 Defense Conversion: Redirecting R&D   OTA

1993 Department of Energy Laboratories: Capabilities and Missions   DOE

1993 Department of Energy Laboratories: A New Partnership with Industry   DOE

REFERENCES

Papers and publications resulting from the National Comparative Research and Development Project are labeled [NCRDP].

Aldrich, H. 1979. *Organizations and Environments*. Englewood Cliffs, N.J.: Prentice-Hall.

Andrews, F., ed. 1979. *Scientific Productivity: The Effectiveness of Research Groups in Six Countries*. Cambridge: New York University Press.

Archibugi, D. 1996. "Review of Nelson (ed.) *National Innovation Systems: A Comparative Analysis*." *Research Policy* 25:838–842.

Bagur, J. D. and A. S. Guissinger. 1987. "Technology Transfer Legislation: An Overview." *Journal of Technology Transfer* 12 (1): 51–63.

Barth, J. and J. Cordes. 1982. "The Economic Recovery Tax Act of 1981: Impacts on R&D, Technological Change, and Innovation." Mimeo.

Battelle Memorial Institute. 1996. *Probable Levels of R&D Expenditures in the U.S.* Columbus, Ohio: Battelle, Columbus Division Resource Management and Economic Analysis Department.

Baxter, J. P. 1946. *Scientists Against Time*. Boston: Little, Brown.

Bernal, J. D. 1953. *Science and Industry in the Nineteenth Century*. Bloomington: Indiana University Press.

Bilstein, R. E. 1989. *Orders of Magnitude: A History of the NACA and NASA, 1915–1990*. Washington, D.C.: NASA, Office of Management, Scientific and Technical Information Division, U.S. Government Printing Office.

Blake, R. and J. Mouton. 1968. *The New Managerial Grid*. Hoston: Gulf Publishing.

Bloch, E. 1991. *Toward a U.S. Technology Strategy*. Washington, D.C.: National Academy Press.

Boundy, R. H. and J. L. Amos, eds. 1990. *A History of the Dow Chemical Physics Lab: The Freedom to be Creative*. New York: Marcel Dekker.

Bozeman, B. 1982. "Organization Structure and Effectiveness of Public Agencies." *International Journal of Public Administration* 4 (3): 235–296.

284   References

——. 1987. *All Organizations Are Public: Binding Public and Private Organization Theory.* San Francisco: Jossey Bass.

——. 1993. "Metricmania and R&D Evaluation." Department of Energy, Evaluation Office, Washington, D.C., October.

——. 1994a. "The Cooperative Technology Paradigm: An Evaluation of U.S. Government Laboratories' Technology Transfer Activities." *Policy Studies Journal* 22 (2).

——. 1994b. "Evaluating Government Technology Transfer: Early Impacts of the 'Cooperative Technology Paradigm.'" Policy Studies Journal 22:322–327. [NCRDP]

——. 1995. "Commercialization of Federal Laboratory Technology." Working paper under review for publication (April).

——. 1996a. "Bureaucratic Red Tape and Formalization: Untangling Conceptual Knots." American Review of Public Administration 25 (1) (March): 1–17.

——. 1996b. "Commercialization of Federal Laboratory Technology." Paper presented at the annual meeting of the American Sociological Association, New York, August. [NCRDP]

——. 1997. "Commercialization of Federal Laboratory Technology: Results of a Study of Industrial Partners." In R. P. Oakey, ed., *New Technology-Based Firms in the 1990s*, 3:127–139. London: Chapman.

Bozeman, B. and S. Bretschneider. 1994. "The 'Publicness Puzzle' in Organization Theory: A Test of Alternative Explanations of Differences Between Public and Private Organizations." *Journal of Public Administration Research and Theory* 4 (April): 197–224.

——. 1995. "Understanding Red Tape and Bureaucratic Delays." In A. Halachmi and G. Bouckaert, eds. *The Enduring Challenges in Public Management*, 81–116. San Francisco: Jossey-Bass.

Bozeman, B. and K. Coker. 1992. "Assessing the Effectiveness of Technology Transfer From U.S. Government R&D Laboratories: The Impact of Market Orientation." *Technovation* 12 (4): 239–255. [NCRDP]

Bozeman, B. and M. Crow. 1988. "U.S. R&D Laboratories and Their Environments: Public and Market Influence." Final Report to the National Science Foundation, Science Resources Section. Syracuse, N.Y.: Technology and Information Policy Program. [NCRDP]

——. 1990. "The Environments of U.S. R&D Laboratories: Political and Market Influences." *Policy Sciences* 23:25–56. [NCRDP]

——. 1991a "Technology Transfer From U.S. Government and University R&D Laboratories." *Technovation* 11 (4) (May): 231–246. [NCRDP]

——. 1991b. "Red Tape and Technology Transfer in Government Labs in the US." *Journal of Technology Transfer* 16 (2) (spring): 29–37. [NCRDP]

——. 1992. "Science Policy: Pork Barrel or Peer Review." *Forum Applied Research and Public Policy* 7:64–73.

——. 1995. *Federal Laboratories in the National Innovation System: Policy Implications of the National Comparative Research and Development Project.* A Report to the U.S. Department of Commerce.

Bozeman, B. and M. Fellows. 1988. "Technology Transfer at the U.S. National

Laboratories: A Framework for Evaluation." *Evaluation and Program Planning* 11:65–75. [NCRDP]

Bozeman, B. and G. Kingsley. In press. "The Research Value Mapping Approach to R&D Assessment." *Journal of Technology Transfer.*

Bozeman, B. and A. Link. 1983. *Investments in Technology: Corporate Strategy and Public Policy Alternatives.* New York: Praeger.

Bozeman, B. and A. Link. 1984 "Tax Incentives for R&D: A Critical Evaluation," *Research Policy* 13 (1) (February): 21–31.

Bozeman, B. and S. Loveless 1987. "Sector Context and Performance: A Comparison of Industrial and Government Research Units." *Administration and Society* 19 (2) (August): 197–235. [NCRDP]

Bozeman, B. and R. McGowan. 1982. "Information Acquisition and Political Environments in Technologically Intensive Government Agencies." In Devendra Sahal, ed., *The Transfer and Utilization of Technical Knowledge.* Lexington, Mass.: Lexington Books.

Bozeman, B. and J. Melkers, eds. 1993. *Evaluating R&D Impacts: Methods and Practice.* Boston: Kluwer Academic Publishers. [NCRDP]

Bozeman, B. and S. Pandey. 1994. "Government Laboratories as a 'Competitive Weapon': Comparing Cooperative R&D in the U.S. and Japan." *Technovation* 14 (3) (April): 145–161. [NCRDP]

——. 1995. "Government Laboratories as a 'Competitive Weapon': A Comparison of U.S. and Japanese Policy." In D. Balkin, ed., *Public Policy and the Management of Innovation in Technology-Based Entrepreneurship,* 231–245. Greenwich, Conn.: JAI Press.

Bozeman, B. and M. Papadakis. 1995a. "Company Interactions with Federal Laboratories: What They Do and Why They Do It." Working paper under review for publication (April).

——. 1996. "Government Laboratories as a 'Competitive Weapon': Comparing Cooperative R&D in the U.S. and Japan." In L. Gomez-Mejia and M. Lawless, eds., *Advances in Global High-Technology Management,* 27–36. Greenwich, Conn.: JAI Press.

——. 1995b. "Firms' Objectives in Industry-Federal Laboratory Technology Development Partnerships." *Journal of Technology Transfer* 20 (3/4) (December): 64–74.

Bozeman, B., M. Papadakis, and K. Coker. 1995a. *Industry Perspectives on Commercial Interactions with Federal Laboratories: Does the Cooperative Technology Paradigm Really Work?* Report to the National Science Foundation, Research on Science and Technology Program, January. [NCRDP]

——. 1995b. *Industry Perspectives on Commercial Interactions with Federal Laboratories: Final Report to the National Science Foundation.* Atlanta, Ga.: School of Public Policy, Georgia Tech.

Bozeman, B., D. Rahm, and M. Crow. 1987. "Technology Transfer to Government and Industry: Who? Why? And to What Effect?" Paper presented at the Harvard Business School Colloquium on Operations and Production Management, November 16. [NCRDP]

Bozeman, B., H. Rainey, and S. Pandey. 1995. "Public and Private Managers'

Perceptions of Red Tape." *Public Administration Review* 55 (6) (December): 567–573.

Bozeman, B. and H. Rainey. In press. "The Bureaucratic Personality Revisited." *American Journal of Political Science.*

Bozeman, B., P. Reed, and P. Scott. 1992. "Red Tape and Task Delays in Public and Private Organizations." *Administration and Society* 24:290–322.

Bozeman, B. and J. Rogers. In press. "Basic Research and Technology Transfer in Federal Laboratories." *Journal of Technology Transfer.*

Bozeman, B. and D. Wittmer. "Technical Roles and Success of Federal Laboratory-Industry Partnerships." Paper presented at the International Meeting of the Society for the Social Study of Science, Bielefeld, Germany, October 1996.

Branscomb, L. M. 1991a. "America's Emerging Technology Policy." Harvard University Working Paper.

——. 1991b. "Toward a U.S. Technology Policy." *Issues in Science and Technology Policy* 7 (4) (summer): 50–55.

——, ed. 1993. *Empowering Technology: Implementing a U.S. Strategy.* Cambridge, Mass.: MIT Press.

Braybrooke, D. and C. E. Lindblom 1963. *A Strategy of Decision: Policy Evaluation as a Social Process.* New York: Free Press of Glencoe.

Brooks, H. 1996. "The Evolution of U.S. Science Policy." In L. Bruce, R. Smith, and C. E. Barfield, eds., *Technology, R&D, and the Economy*, 15–41. Washington, D.C.: The Brookings Institution and American Enterprise Institute.

Brooks, H. and R. Schmitt 1985. "Current Science and Technology Policy Issues: Two Perspectives." Occasional Paper no. 1, George Washington University.

Browning, G. 1995. "Tense Days Down in the Lab." *National Journal* 27 (April 22): 1005.

Brownstein, C. 1996. *Hearing on NIST Research Programs.* Technology Subcommittee of the House Science Committee, June 25.

Bucy, J. F. 1985. "Meeting the Competitive Challenge: The Case for R&D Tax Credits." *Issues in Science and Technology* (summer): 69–78.

Bush, V. 1945. *Science: The Endless Frontier. A Report to the President on a Program for Postwar Scientific Research.* Washington, D.C.: Office of Scientific Research and Development.

Callon, M. et al. 1992. "The Management and Evaluation of Technological and the Dynamics of Techno-Economic Networks: The Case of the AFME." *Research Policy* 21 (3): 215–236.

Carlson, W. B. 1991. *Innovation as a Social Process: Elihu Thomson and the Rise of General Electric, 1870–1900.* Cambridge, New York: Cambridge University Press.

Cattell Press. 1970. *Industrial Research Laboratories of the United States.* 13th ed. New York: R.R. Bowker.

——. 1983. *Industrial Research Laboratories of the United States.* 18th ed. New York: R.R. Bowker.

Chapman, Richard L. 1986. *The Uncounted Benefits: Federal Efforts in Domestic Technology Transfer.* Denver: Denver Research Institute, 1986.

Cheng, J. and B. Bozeman. 1993. "Resource Dependence and Interorganizational

Linkage Among R&D Labs: The Impact of Research Orientations." *Journal of High Technology Management Research* 4 (2) (fall): 255–273. [NCRDP]

Cho, M. H. 1991. "Cooperative R&D Arrangements: Impacts on United States R&D Laboratories' Research Effectiveness." Ph.D. diss., Syracuse University. [NCRDP]

Choi, Y.-H. 1996. "Partnering Government Laboratories with Industry: A Comparison of the U.S. and Japan." Ph.D. diss., Syracuse University. [NCRDP]

Clinton-Gore. 1992. "Technology: The Engine of Economic Growth. A National Technology Policy for America." Contact: Ellis Mottur, Clinton-Gore National Campaign Headquarters, November 1.

Coker, K. 1994. "Federal Laboratories as Industry Partners (Technology Transfer)." Ph.D. diss., Syracuse University.

Collins, D. 1990. *The Story of Kodak.* New York: Abrams.

Conference Board. 1987. *Getting More Out of Science and Technology.* Research Report No. 904. New York: Conference Board.

Council on Competitiveness. 1993. *Industry as a Customer of the Federal Laboratories.* Washington, D.C.: U.S. Government Printing Office.

Coursey, D. and B. Bozeman. 1988. "A Typology of Industry-Government Laboratory Cooperative Research: Implications for Government Laboratory Policies and Competitiveness." In A. Link and G. Tassey, eds., *Cooperative Research.* Boston: Kluwer Publishing. [NCRDP]

——. 1992. Technology Transfer in U.S. Government and University Laboratories: Advantages and Disadvantages for Participating Laboratories." *IEEE Transactions on Engineering Management* 39 (4): 347–351. [NCRDP]

Crease, R. P. 1993. "The National Labs and Their Future." *Forum for Applied Research and Public Policy* 8 (4): 96–104.

Crow, M. 1985. *The Effect of Publicness on Organizational Performance: A Comparative Analysis of R&D Laboratories.* Springfield, Va.: NTIS (PB85–216646). [NCRDP]

——. 1987. "Synthetic Fuel Technology Non-Development and the Hiatus Effect: The Implications of Inconsistent Public Policy." In E. J. Yanarella and W. Green, eds., *The Unfulfilled Promise of Synthetic Fuels: Technological Failure, Policy Immobilism, or Commercial Illusion?* 33–52. Westport, Conn.: Greenwood Press.

——. 1988. "Technology and Knowledge Transfer in Energy R&D Laboratories: An Analysis of Effectiveness." *Evaluation and Program Planning* 11:85–95. [NCRDP]

——. 1989. "Technology Development in Japan and the United States: Lessons from the High-Technology Superconductivity Race." *Science and Public Policy* 16:322–344.

Crow, M. and B. Bozeman. 1987a. "R&D Laboratories' Environmental Contexts: Are the Government Lab-Industrial Lab Stereotypes Still Valid?" *Research Policy* 13 (December): 329–355.

——. 1987b. "A New Typology for R&D Laboratories: Implications for Policy Analysis." *Journal of Policy Analysis and Management* 6 (3): 328–341. [NCRDP]

———. 1989a. "Bureaucratization in the Laboratory." *Research Technology Management* 32 (5) (September/October): 30–32.

———. 1989b. "Information Products and Policies of Federal Laboratories." In C. McClure and P. Hernon, eds., *Perspectives on U.S. Government Scientific and Technological Information Policies*, 193–215. Norwood, N.J.: Ablex Publishing. [NCRDP]

———. 1991. "R&D Laboratories in the USA: Structure, Capacity, and Context." *Science and Public Policy* 18 (3): 165–179. [NCRDP]

Crow, M. and M. Emmert. 1987. "Organized Science and Public-Private Cooperation." *Journal of Management* 13 (1) (summer): 55–67.

Crow, M. and S. Nath. 1990. "Technology Strategy Development in Japanese Industry: An Assessment of Market and Government Influences." *Technovation* 10 (5) (July): 333–346.

———. 1992. "Technology Strategy Development in Korean Industry: An Assessment of Market and Government Influences." *Technovation* 12 (2) (March): 119–136.

Crow, M., M. Emmert, and C. Jacobson. 1991. "Government-Supported Industrial Research Institutes in the United States." *Policy Studies Journal* 19 (1) (fall): 59–74.

Dahl, R. and C. Lindblom. 1953. *Politics, Economics, and Welfare.* New York: Harper and Row.

Dalton, D. R. et al. 1980. "Organizational Structure and Performance: A Critical Review." *Academy of Management Review* 5:49–60.

Danhof, C. H. 1968. *Government Contracting and Technological Change.* Westport, Conn.: Greenwood Press.

Daniels, G. H. 1971. *Science in American Society.* New York: Knopf.

De la Barre, D. M. 1986. "Federal Technology Transfer Act of 1986: PL99–502 at a Glance." *The Journal of Technology Transfer* 11 (1): 19–20.

Dosi, G., C. Freeman, R. Nelson, and L. Soete, eds. 1988. *Technical Change and Economic Theory.* London: Pinter Publishers.

Dupree, A. H. 1957. *Science in the Federal Government: A History of Policies and Activities to 1940.* Cambridge, Mass.: The Belknap Press of Harvard University Press.

———. 1986. *Science in the Federal Government.* Baltimore: Johns Hopkins University Press.

Eddy, E. D., Jr. 1957. *Colleges for Our Land and Time: The Land-Grant Idea in American Education.* New York: Harper.

Eisenberg, R. 1995. "Public Research and Private Development: Patents and Technology Transfer in the Human Genome Project." Draft Manuscript.

Eisner, R. 1985. "R&D Tax Credits: A Flawed Tool." *Issues in Science and Technology* (summer): 79–86.

Emmert, M. and M. Crow. 1988. "Public, Private, and Hybrid Organizations: An Empirical Examination of the Role of Publicness." *Administration and Society* 20:216–44.

England, J. M. 1983. *A Patron for Pure Science: The National Science Foundation's Formative Years, 1945–57.* Washington, D.C.: National Science Foundation.

Falcone, S. 1991. "Structure, Sector, Technology, and Size in Research and Development Laboratories." Ph.D. diss., Syracuse University. [NCRDP]

Ford, J. and J. Slocum. 1977. "Size, Technology, Environment, and the Structure of Organizations." *Academy of Management Review* 2:561–575.

Francis, S. 1996. "Warhead Politics: Livermore and the Competitive System of Nuclear Weapon Design (Los Alamos National Laboratory, New Mexico, Lawrence Livermore National Laboratory, California)." Ph.D. diss., Massachusetts Institute of Technology.

Freeman, C. 1995. "The 'National System of Innovation' in Historical Perspective." *Cambridge Journal of Economics* 19:5–24.

Fusfeld, H. 1986. *The Technical Enterprise.* Cambridge, Mass.: Ballinger.

Geiger, R. L. 1986. *To Advance Knowledge: The Growth of American Research Universities, 1900–1940.* New York: Oxford University Press.

——. 1992. "The Dynamics of University Research in the United States: 1945–1990." In T. G. Whiston and R. L. Geiger, eds., *Research and Higher Education: The United Kingdom and the United States,* 3–17. Buckingham, U.K.: Open University Press.

——. 1993. *Research and Relevant Knowledge: American Research Universities Since World War II.* New York: Oxford University Press.

Gillespie, G. C. 1988. "Federal Laboratories: Economic Development and Intellectual Property Constraints." *Journal of Technology Transfer* 13 (1): 20–26.

Goodman, R., S. Brownlee, and T. Watson. 1995. "Should the Labs get Hit?" *U.S. News and World Report* 119, November 6, p. 83.

Graham, H. D. and N. Diamond. 1997. *The Rise of American Research Universities: Elites and Challengers in the Postwar Era.* Baltimore: Johns Hopkins University Press.

Graham, M. B. W. and B. H. Pruitt. 1990. *R&D for Industry: A Century of Technical Innovation at Alcoa.* New York: Cambridge University Press.

Gummett, P. and M. Gibbons. 1978. "Government Research for Industry: Recent British Developments." *Research Policy* 7:268–290.

Guralnick, S. M. 1979. "The American Scientist in Higher Education, 1820–1910." In N. Reingold, ed., *Sciences in the American Context: New Perspectives,* 99–142. Washington, D.C.: Smithsonian Institution Press.

Haber, L. F. 1958. *The Chemical Industry During the Nineteenth Century: A Study of the Economic Aspect of Applied Chemistry in Europe and North America.* Oxford: Clarendon Press.

Hager, M. 1996. "The War on Washington Waste." *Consumer Digest* (September/October): 54–60.

Hammond, J. W. 1941. *Men and Volts: The Story of General Electric.* Philadelphia: Lippincott.

Hanson, D. 1992. "Federal Research Labs Criticized for Lax Accounting." *Chemical and Engineering News* 70 (July 27): 26.

Hedges, S. 1990. "Messing up the Nuclear Cleanup." *U.S. News and World Report* 108, February 19, p. 26.

Henderson, W. O. 1983. *Friedrich List: Economist and Visionary, 1789–1846.* London: Frank Cass.

Herrmann, J. F. 1983. "Redefining the Federal Government's Role in Technology Transfer." *Research Management* 26 (1) (January/ February): 21–24.

Hewlett, R. G. and O. E. Anderson. 1962. *A History of the United States Atomic Energy Commission.* University Park: Pennsylvania State University Press.

Hounshell, D. A. 1984. *From the American System to Mass Production, 1800–1932: The Development of Manufacturing Technology in the United States.* Baltimore: Johns Hopkins University Press.

Hounshell, D. A. and J. K. Smith. 1988. *Science and Corporate Strategy: Du Pont R&D, 1902–1980.* New York: Cambridge University Press.

Hughes, T. P. 1983. *Networks of Power: Electrification in Western Society, 1880–1930.* Baltimore: Johns Hopkins University Press.

Ihde, A. J. 1964. *The Development of Modern Chemistry.* New York: Harper and Row.

Irvine, J., B. Martin, M. Schwartz, and K. Pavitt. 1981. "Government Support for Industrial Research in Norway." Science Policy Research Unit; University of Sussex, July, 197–336.

Jaffe, A. and M. Trajtenberg. 1996. *NBER Working Paper 5712: Flows of Knowledge from Universities and Federal Labs: Modeling the Flow of Patent Citations Over Time and Across Institutional and Geographic Boundaries.* Cambridge: National Bureau of Economic Research.

Jaffe, A., M. Fogarty, and B. Banks, 1997. *NBER Working Paper 6044: Evidence from Patents and Patent Citations on the Impact of NASA and Other Federal Labs on Commercial Innovation.* Cambridge: National Bureau of Economic Research.

Johnson, C. 1984. *The Industrial Policy Debate.* San Francisco: ICS Press.

Joly, P. B. and V. Mangematin. 1996. "Profile of Public Laboratories, Industrial Partnerships, and Organization of R&D: The Dynamics of Industrial Relationships in a Large Organization." *Research Policy* 25 (6): 901–922.

Kamien, M. and N. Schwartz. 1982. *Market Structure and Innovation.* New York: Cambridge University Press.

Kash, D. 1989. *Perpetual Innovation: The New World of Competition.* New York: Basic Books.

Kash, D. and R. Rycroft. 1993. "Nurturing Winners with Federal R&D." *Technology Review* 96 (8): 58–74.

———. 1994. "Technology Policy: Fitting Concept with Reality." *Technological Forecasting and Social Change* 47 (1): 35–48.

———. 1995. "U.S. Federal Government R&D and Commercialization: Can't Get There from Here." *R&D Management* 25 (1): 71–89.

Kearns, D. 1990. "Federal Labs Teem with R&D Opportunities." *Chemical Engineering* 97 (4): 131–137.

Keyworth, G. A. 1983. "Federal R&D and Industrial Policy." *Science* 220: 1122–1125.

Kimberly, J. 1976. "Organizational Size and the Structuralist Perspective: A Review, Critique, and Proposal." *Administrative Science Quarterly* 21 (4): 571–597.

Kingsley, G. 1993. "The Use of Case Studies in R&D Impact Evaluations." In B. Bozeman and J. Melkers, eds., *Evaluating R&D Impacts: Methods and Practice.* Boston: Kluwer Academic Publishers.

Kingsley, G. 1994. "Leading the Horse to Water: Evaluating Technology Transfer and Technology Absorption from State-Sponsored Research, Development, and Demonstration Projects." Ph.D. diss., Syracuse University. [NCRDP]

Kingsley, G. and B. Bozeman. 1992. "Bureaucratization and Public-Private Organizations: The Impact of Task, Structure, and External Environment." Paper presented at joint meeting of Institute for Management Science/Operations Research Society of America, Orlando, Florida, April.

——. 1996. "Commercial Interactions with Federal Laboratories." *Materials Technology*, forthcoming.

——. 1997. "Charting the Routes to Commercialization: The Absorption and Transfer of Energy Conservation Technologies." *International Journal of Global Energy Issues* 9 (1/2): 8–15.

Kingsley, G., B. Bozeman, and K. Coker. 1995. "Technology Transfer and Absorption: An 'R&D Value-mapping' Approach to Evaluation," *Research Policy* 25 (6): 967–995. [NCRDP]

Krieger, J. H. 1987. "Cooperation Key to US Technology Remaining Competitive." *Chemical and Engineering News* (April 27).

Lan, Z. 1991. "The Impact of Resource Dependence Patterns on University Research and Development Labs: An Exploratory Study." Ph.D. diss., Syracuse University. [NCRDP]

Landau, R. and N. B. Hannay, eds. 1981. *Taxation, Technology, and the U.S. Economy.* New York: Pergamon.

Latour, B. 1979. *Laboratory Life: The Social Construction of Scientific Facts.* Beverly Hills: Sage Publications.

Lepkowski, W. 1991. "Technology Transfer Problems Discussed: John Dingell Investigates Why So Little Transfer of Technology from Federal Laboratories to Industry." *Chemical & Engineering News* 69 (August 5): 13.

Leslie, S. W. 1993. *The Cold War and American Science: The Military-Industrial-Academic Complex at MIT and Stanford.* New York: Columbia University Press.

Leyden, D., A. Link, and B. Bozeman. 1989. "The Effects of Governmental Financing on Firms' R&D Activities: A Theoretical and Empirical Investigation." *Technovation* 9:561–575. [NCRDP]

Link, A. 1987. "Public Policy for Basic Research." Paper presented at the National Science Foundation Workshop on Science Policy (September).

——. 1996. *Evaluating Public Sector Research and Development.* Westport, Conn.: Praeger.

Link, A. and L. Bauer. 1989. *Cooperative Research in U.S. Manufacturing: Assessing Policy Initiatives and Corporate Strategies.* Lexington, Mass.: Lexington Books.

Link, A. and B. Bozeman. 1991. "Innovative Behavior in Small-Sized Firms." *Small Business Economics* 3:179–184. [NCRDP]

Link, A. and G. Tassey. 1987. *Strategies for Technology-Based Competition.* Lexington, Mass.: Heath.

Link, A., B. Bozeman, and M. Crow. 1988. "The Effect of Government Funding on Innovative Activity in Industrial Research Laboratories." *Proceedings of the International Meetings on Cybernetics and Systems Research.* Vienna, Austria. [NCRDP]

——. 1989. "Federal Funding and R&D Output in U.S. Industrial Laboratories." *IEEE Transactions* 36 (2) (May). [NCRDP]

Lippman, T. 1989. "Engineer Fired After Calling Nuclear Plant Vulnerable." *Washington Post* 113, December 9, p. 3.

List, F. 1841. *The National System of Political Economy*. English ed. 1904. London: Longmans.

Loveless, S. 1985. "Sector Status, Structure, and Performance: A Comparison of Public and Private Research Units." Ph.D. diss., Syracuse University.

Lundvall, B., ed. 1992. *National Systems of Innovation: Towards a Theory of Innovation and Interactive Learning*. London: Pinter Publishers.

MIT Commission on Industrial Productivity. 1989. *Made in America: Regaining the Competitive Edge*. Cambridge: MIT Press.

Marcson, S. 1960. *The Scientist in American Industry: Some Organizational Determinants in Manpower Utilization*. New York: Published in cooperation with the Industrial Relations Section, Dept. of Economics, Princeton University: Harper.

Markusen, A., J. Raffel, M. Oden, and M. Llanes. 1995. *Coming in from the Cold: The Future of Los Alamos and Sandia National Laboratories*. Pascataway, N.Y.: Center for Urban Policy Research, Rutgers University.

Mayfield, L. and E. Schutzman. 1987. "Status Report on the NSF Engineering Research Centers Program." *Research Management* 30 (1) (January/February): 35–41.

McKelvey, B. 1982. *Organizational Systematics*. Berkeley: University of California Press.

McKelvey, M. 1991. "How Do National Systems of Innovation Differ? A Critical Analysis of Porter, Freeman, Lundvall, and Nelson." In G. Hodgson and E. Screpanti, eds., *Rethinking Economics: Markets, Technology, and Economic Evolution*. Aldershot: Edward Elgar.

McMahon, A. M. 1984. *The Making of a Profession: A Century of Electrical Engineering in America*. New York: IEEE Press.

Meyer, M. and Associates. 1977. *Organizations Environments*. San Francisco: Jossey-Bass.

Morella, C. 1996. "Oversight Review of the Research Laboratory Programs at the National Institute of Standards and Technology." *Introductory Remarks by the Chairwoman of the Technology Subcommittee of the House Science Committee*. June 25.

Mortenson, L. E. 1991. "Management of Inter-organizational Policy Conflict: A Study of Mission and Policy Conflict Between Two DHHS Agencies' Interorganizational Policy." Ph.D. diss., University of Southern California. [NCRDP]

Mowery, D. C. 1981. "The Emergence and Growth of Industrial Research in American Manufacturing, 1899–1946." Ph.D diss., Stanford University.

——. 1992. "The U.S. National Innovation System: Origins and Prospects for Change." *Research Policy* 21 (2) (April): 125.

Mowery, D. C. and N. Rosenberg. 1989. *Technology and the Pursuit of Economic Growth*. New York: Cambridge University Press.

——. 1993. "The U.S. National Innovation System." In R. R. Nelson, ed., *National*

Innovation Systems: A Comparative Analysis, 29–75. New York: Oxford University Press.

Mowery, D. C., N. Rosenberg, R. Landau. 1992. *Technology and the Wealth of Nations*. Stanford: Stanford University Press

National Academy of Sciences. 1978. *Technology, Trade, and the U.S. Economy*. Washington, D.C.: National Academy of Sciences.

——. 1986. *New Alliances and Partnerships in American Science and Engineering*. Washington, D.C.: National Academy Press.

National Cooperative Research Act of 1984.

National Governor's Association. 1987. *The Role of Science and Technology in Economic Competitiveness*. New York: National Governor's Association.

National Science Foundation. 1987. *National Patterns of R&D Resources, Funds, and Manpower in the US, 1953–1976*. Washington, D.C.: National Science Foundation.

——. 1990. *Science and Engineering Personnel: A National Overview*. Washington, D.C.: U.S. Government Printing Office.

——. 1991. *Science and Engineering Indicators*. Washington, D.C.: U.S. Government Printing Office (also 1993 and 1996).

——. 1993. *Characteristics of Doctoral Scientists and Engineers in the United States*. Washington, D.C.: U.S. Government Printing Office.

Nelson, R. R. 1962. "The Link Between Science and Invention: The Case of the Transistor." In R. Nelson, ed., *The Rate and Direction of Inventive Activity*. Princeton: Princeton University Press for the National Bureau of Economic Research.

——. 1984. *High Technology Policies: A Five Nation Comparison*. Washington, D.C.: American Enterprise Institute for Public Policy Research.

——. 1992. "National Innovation Systems: A Retrospective on a Study." *Industrial and Corporate Change* 1:347–374.

——. 1996. "The Agenda for Growth Theory: A Different Point of View." Draft manuscript.

——, ed. 1993. *National Innovation Systems: A Comparative Analysis*. New York: Oxford University Press.

Nelson, R. R., M. J. Peck, and E. D. Kalachek. 1967. *Technology, Economic Growth, and Public Policy: A Rand Corporation and Brookings Institution Study*. Washington, D.C.: Brookings Institution.

Niosi, J. 1988. *The Decline of the American Economy*. New York: Black Rose Books.

——. 1991. *Technology and National Competitiveness: Oligopoly, Technological Innovation, and International Competition*. Montreal: McGill-Queen's University Press.

——. 1993. "Strategic Partnerships in Canadian Advanced Materials." *R&D Management* 23 (1): 17–27.

——. 1994. *New Technology Policy and Technical Innovations in the Firm*. Londres: Pinter/Leicester University Press.

Niosi, J. and M. Bergeron. 1992. "Technical Alliances in the Canadian Electronics Industry: An Empirical Analysis." *Technovation* 12 (5): 309–322.

Niosi, J., P. Saviotti, B. Bellon, and M. Crow. 1993. "National Systems of Innovation: In Search of a Workable Concept." *Technology in Society* 15:207–227.

Norris, W. 1985. "Cooperative R&D: A Regional Strategy." *Issues in Science and Technology* 1 (2): 92–102.

Office of Naval Research. 1987. *Office of Naval Research: Forty Years of Excellence in Support of Naval Science, 40th Anniversary, 1946–1986.* Arlington, Va.: Office of Naval Research, Dept. of the Navy.

Office of Technology Assessment. 1993. *Defense Conversion: Redirecting R&D.* Washington, D.C.: U.S. Government Printing Office.

Orlans, H. 1967. *Contracting for Atoms: A Study of Public Policy Issues Posed by the Atomic Energy Commission's Contracting for Research, Development, and Managerial Services.* Washington, D.C.: Brookings Institution.

Orsenigo, L. 1989. *The Emergence of Biotechnology: Institutions and Markets in Industrial Innovation.* London: Pinter.

Pandey, S. and B. Bozeman. 1994. "Government Laboratories as a Competitive Weapon: Comparing Cooperative R&D in the U.S. and Japan." In D. Balkin, ed., *Public Policy and the Management of Innovation in Technology-Based Entrepreneurship.* Greenwich, Conn.: JAI Press. [NCRDP]

Papadakis, M. 1992. "Federal Laboratory Missions, 'Products,' and Competitiveness." *Journal of Technology Transfer* 17 (2): 47–53. [NCRDP]

——. 1994. "Did (or Does) the United States Have a Competitiveness Crisis?" *Journal of Policy Analysis and Management* 13 (1): 1–20. [NCRDP]

Papadakis, M. et al. 1995. "The Japanese Government Laboratory System." *Japan Technical Journal* 5 (1): 18–39.

Patel, P. and K. Pavitt. 1994. "National Systems of Innovation: Why They are Important and How They Might be Measured and Compared." *Economics of Innovation and New Technology* 3:77–95.

Patent and Trademark Laws Amendment, 1980.

Pavitt, K. 1984. "Sectoral Patterns of Technical Change: Towards a Taxonomy and a Theory." *Research Policy* 13 (6): 343–375.

——. 1995 "Review of Lundvall (ed.) *National Systems of Innovation,*" *Research Policy* 24:320.

Peck, M. 1986. "Joint R&D: The Case of Microelectronics and Computer Technology Corporation." *Research Policy* 15 (5): 219–223.

Perry, J. L. and H. Rainey. 1988. "The Public-Private Distinction in Organization Theory." *Academy of Management Review* 13 (2): 182–201.

Pondy, L. 1969. "Effects of Size, Complexity, and Ownership on Administrative Intensity." *Administrative Science Quarterly* 14 (1): 47–60.

Porter, M. 1990. *The Competitive Advantage of Nations.* London: McMillan.

President's Commission on Industrial Competitiveness. 1985. *Global Competition: The New Reality.* Washington, D.C.: U.S. Government Printing Office.

Rahm, D. 1989, "The Influence of Direct Funding, Indirect Funding, and Institutional Design Policies on Industrial Research and Development (Funding, R&D)." Ph.D. diss., Syracuse University. [NCRDP]

Rahm, D., B. Bozeman, and M. Crow. 1988. "Domestic Technology Transfer and Competitiveness: An Empirical Assessment of Roles of University and Government R&D Laboratories." *Public Administration Review* 48 (6) (November/December): 969–978. [NCRDP]

Rainey, H. G., S. Pandey, and B. Bozeman. 1995. "Research Note: Public and Private Managers' Perceptions of Red Tape." *Public Administration Review* 55 (November/December): 567–574.

Radetsky, Peter. 1997. "The Gulf War Within." *Discover* 8 (18): 68.

Reich, L. S. 1985. *The Making of American Industrial Research: Science and Business at GE and Bell, 1876–1926.* New York: Cambridge University Press.

Ricketts, P. C. 1934. *History of Rensselaer Polytechnic Institute, 1824–1934.* 3d ed. New York: Wiley.

Roessner, J. D. and A. Bean. 1991. "How Industry Interacts with Federal Laboratories." *Research Technology Management* 34 (4) (July/August): 22.

Roessner, J. D. and A. S. Bean. 1993. "Industry Interaction with Federal Lab Pays Off." *Research Technology Management* 36 (5): 38–40.

Rosenberg, C. E. 1977. "Rationalization and Reality in Shaping American Agricultural Research, 1875–1914." *Social Studies of Science* 7 (November): 401–422.

Rosenberg, N. 1963. "Technical Change in the Machine Tool Industry, 1840–1910." *Journal of Economic History* 23:414–443.

——, ed. 1969. *American System of Manufactures.* Edinburgh: Edinburgh University Press.

Rosenberg, N. and R. R. Nelson. 1994. "American Universities and Technical Advance in Industry." *Research Policy* 23:323–348.

Rosenberg, N., A. C. Gelijns, and H. Dawkins. 1995. *Sources of Medical Technology: Universities and Industry.* Washington, D.C.: National Academy Press.

Rosenbloom, R. and W. Spencer. 1996. *Engines of Innovation: U.S. Industrial Research at the End of an Era.* Boston: Harvard Business School Press.

Ross, E. D. 1942. *Democracy's College: The Land-Grant Movement in the Formative Stage.* Ames: Iowa State College Press.

Roy, R. and D. Shapley. 1985. *Lost at the Frontier: U.S. Science and Technology Policy Adrift.* Philadelphia: ISI Press.

Rudolph, F. 1962. The American College and University: A History. New York: Vintage Books.

Rycroft, R. and D. Kash. 1994. "Complex Technology and Community: Implications for Policy and Social Science." *Research Policy* 23 (6): 613–626.

Saddler, J. 1992. "Technology Transfer: Labs Lag in Turning Research into Products." *Wall Street Journal,* December 10, B2.

Sapolsky, H. 1979. "Academic Science and the Military: The Years Since the Second World War." In N. Reingold, ed., *The Sciences in the American Context: New Perspectives,* 379–399. Washington, D.C.: Smithsonian Institution Press.

——. 1990. *Science and the Navy: The History of the Office of Naval Research.* Princeton: Princeton University Press.

Schriesheim, A. 1990. "Toward a Golden Age for Technology Transfer." *Issues in Science and Technology* (winter): 52–58.

——. 1996. "Obstacles to 21st Century Technology." Mid-America Regulatory Commissioners Conference, Chicago, June 17.

Scott, P. 1994. "The Effects of Accountability, Responsiveness, and Professionalism on Bureaucratic Discretion: An Experiment in Street-level Decision-making." Ph.D. diss., Syracuse University.

Servos, J. W. 1990. *Physical Chemistry from Ostwald to Pauling: The Making of Science in America.* Princeton: Princeton University Press.

Simons, G. R. 1993. "Industrial Extension and Innovation." In Lewis Branscomb, ed., *Empowering Technology: Implementing a U.S. Strategy,* 167–202. Cambridge: MIT Press.

Smilor, R. and D. Gibson, 1991. "Accelerating Technology Transfer in R&D Consortia." *Research Technology Management* 34 (1) (January/February): 44.

Smith, B. L. R. 1990. *American Science Policy Since World War II.* Washington, D.C.: Brookings Institution.

Smith, E. T. 1993. "Can Cold Wars Keep the Home Fires Burning," *Business Week* 115 (June 21): 1.

Snow, C. and L. Hrebiniak. 1980. "Strategy, Distinctive Competence, and Organizational Performance." *Administrative Science Quarterly* 25 (2): 317–336.

Stark, E. 1984. "The Federal Laboratories: Technology Resources and Transfer Champions." Paper presented at American Chemical Society, Philadelphia, PA, August.

Stewart, I. 1948. *Organizing Scientific Research for War: The Administrative History of the Office of Scientific Research and Development.* Boston: Little, Brown.

Tarter, C. B. 1996. *The Department of Energy's Budget Request for 1997.* Hearing of the Subcommittee on Strategic Forces of the Senate Committee on Armed Services, March 13.

Teich, A. and W. Lambright. 1976. "The Redirection of a Large National Laboratory." *Minerva* 14 (4): 447–474

Thackray, A., J. L. Sturchio, P. T. Carroll, and R. Bud. 1985. *Chemistry in America, 1876–1976: Historical Indicators.* Boston: Kluwer Academic Publishers.

Thiesmeyer, L. R. and J. E. Burchard. 1947. *Combat Scientists.* Boston: Little, Brown.

Tolbert, P. 1985. "Institutional Environments and Resource Dependence: Sources of Administrative Structure in Institutions of Higher Education." *Administrative Science Quarterly* 30 (1): 1–13.

Toren, N. 1978a. "The Determinants of the Potential Effectiveness of Government-Supported Industrial Research Institutes." *Research Policy* 7:362–382.

———. 1978b. "The Structure and Management of Government Research Institutes: Some Problems and Suggestions." *R&D Management* 7:5–10.

Trescott, M. M. 1981. *The Rise of the American Electrochemicals Industry, 1880–1910: Studies in the American Technological Environment.* Westport, Conn.: Greenwood Press.

U.S. Department of Energy. 1975. *Report of the Field and Laboratory Utilization Study Group.* Washington, D.C.: U.S. Government Printing Office.

———. 1978. *Report to Congress: The Multiprogram Laboratories: A National Resource for Nonnuclear Energy Research, Development and Demonstration.* Washington, D.C.: U.S. Government Printing Office.

———. 1981. *Federal Energy R&D Priorities.* Washington, D.C.: U.S. Government Printing Office.

———. 1982. *The Department of Energy Multiprogram Laboratories: The Final Report of the Multiprogram Laboratory Panel.* A Report of the Energy Research Advisory Board. Washington, D.C.: U.S. Government Printing Office.

———. 1983. *The Federal Role in Energy Research and Development.* A Report of the

Energy Research Advisory Board. Washington, D.C.: U.S. Government Printing Office.

——. 1985. *Guidelines for DOE Long Term Civilian Research and Development.* Washington, D.C.: U.S. Government Printing Office.

——. 1986. *The Coordination of Long-Term Energy Research and Development Planning.* Washington, D.C.: U.S. Government Printing Office.

——. 1990. *The Future Strategic Role of the Department of Energy's Multiprogram Research Laboratories.* Washington, D.C.: U.S. Government Printing Office.

——. 1991. *Nuclear Weapons Complex Reconfiguration Study.* Washington, D.C.: U.S. Government Printing Office.

——. 1992. *Report to the Secretary on the DOE National Laboratories.* Washington, D.C.: U.S. Government Printing Office.

——. 1993. *Department of Energy Laboratories: Capabilities and Missions.* Washington, D.C.: U.S. Government Printing Office.

——. 1994a. *Changes and Challenges at the Department of Energy Laboratories.* Washington, D.C.: U.S. Government Printing Office.

——. 1994b. *DOE's National Laboratories: Adopting New Missions and Managing Effectively Pose Significant Challenges.* Washington, D.C.: U.S. Government Printing Office.

——. 1994c. *The National Laboratories of the United Kingdom, France, Germany in Transition: Implications for the Department of Energy Laboratory System.* Washington, D.C.: U.S. Government Printing Office.

——. 1994d. Secretary of Energy Advisory Board. *Task Force on Alternative Futures for the Department of Energy National Laboratories.* Washington, D.C.: U.S. Government Printing Office.

——. 1996. *Strategic Laboratory Missions Plan—Phase 1.* Washington, D.C.: U.S. Government Printing Office.

U.S. General Accounting Office. 1989. *Technology Transfer: Implementation Status of the Federal Technology Transfer Act of 1986.* Washington, D.C.: U.S. Government Printing Office.

U.S. House of Representatives Committee on Science. 1987. *Report of the Science Policy Task Force.*

U.S. Office of Technology Assessment. 1993 *Defense Conversion: Redirecting R&D.* Washington, D.C.: U.S. Government Printing Office.

Veysey, L. R. 1965. *The Emergence of the American University.* Chicago: University of Chicago Press.

Vinck, D., B. Kahane, P. Laredo, and J. Meyer. 1993. "A Network Approach to Studying Research Programmes: Mobilizing and Coordinating Public Responses to HIV/AIDS." *Technology Analysis & Strategic Management* 5 (1): 39–54.

Wasserman, N. H. 1985. *From Invention to Innovation: Long-distance Telephone Transmission at the Turn of the Century.* Baltimore: Johns Hopkins University Press.

White House Science Council. 1983. *Report of The Federal Laboratory Review Panel Report.* (Packard Report). Washington D.C.: U.S. Government Printing Office.

Wiebe, R. H. 1967. *The Search for Order, 1877–1920.* New York: Hill and Wang.

Wise, G. 1985. *Willis R. Whitney, General Electric, and the Origins of U.S. Industrial Research.* New York: Columbia University Press.